Charled Romley Alder Wright

Metals and their chief industrial Applications

Being, with some considerable Additions

Charled Romley Alder Wright

Metals and their chief industrial Applications
Being, with some considerable Additions

ISBN/EAN: 9783337069377

Printed in Europe, USA, Canada, Australia, Japan

Cover: Foto ©ninafisch / pixelio.de

More available books at **www.hansebooks.com**

METALS

AND THEIR

CHIEF INDUSTRIAL APPLICATIONS.

BEING, WITH SOME CONSIDERABLE ADDITIONS,
THE SUBSTANCE OF A COURSE OF LECTURES DELIVERED AT
THE ROYAL INSTITUTION OF GREAT BRITAIN IN 1877.

BY

CHARLES R. ALDER WRIGHT, D.Sc., &c.,

*Lecturer on Chemistry
in
St. Mary's Hospital Medical School.*

London:
MACMILLAN AND CO.
1878.

[*The Right of Translation and Reproduction is Reserved.*]

LONDON:
R. CLAY, SONS, AND TAYLOR, PRINTERS,
BREAD STREET HILL,

CONTENTS.

CHAPTER I.

METALS AND THEIR NATURAL SOURCES.

SECT.		PAGE
1.	Distinction between elements and compounds and between metals and non-metals	1
2.	List of elements	3
3.	Metals of greater technical importance	4
4.	Characteristics of metals	5
5.	Native metals and ores: characters of chemical actions involved in metallurgy	7
6.	Classification of metal-extracting processes . . .	8
7.	General chemical characters of metal-extracting processes	11
8.	Native metals	12
9.	Simple ores	13
10.	Complex ores	15
11.	Relationships between heat-disturbance and chemical reactions in wet processes	16
12.	Relationships between laws connecting heat-disturbance and temperature of initial action	17
13.	Relationships between laws connecting heat-disturbance and temperature of initial action	21
14.	Reciprocal reactions	23

CONTENTS.

CHAPTER II.

METALLURGY OF THE PRECIOUS OR NOBLE METALS.

SECT.		PAGE
15. GOLD : Gold-washing	. . . . . .	24
16. Amalgamation process for working gold quartz	. .	27
17. Refining of gold	. . . .	28
18. Wet processes for gold extraction	. . . .	29
19. SILVER : Extraction by Lead processes ; cupellation	.	30
20. Liquation ; Pattinsonage ; Parkes' process	. . .	35
21. Wet processes for silver-extraction	. . . .	37
22. Patio process of amalgamation	. . . . .	39
23. Saxon process of amalgamation	. . . . .	41
24. Silver refining : quartation	. . . . . .	42
25. PLATINUM	. . . .	43
26. Wollaston's process ; oxyhydrogen furnace	. . .	45
27. MERCURY : Aludel process	. . . . . .	49
28. Palatinate gallery	. . . .	51

CHAPTER III.

METALLURGY OF THE MORE IMPORTANT BASE (READILY OXIDIZABLE) METALS.

29. IRON : Characters of chief ores	. . . . .	53
30. Direct and Indirect processes; steel, pig-iron, and wrought-iron.	. . . .	55
31. Catalan forge ; Siemens' processes	. . .	56
32. Blast furnace	. . . .	61
33. Chemical changes taking place in the blast furnace	.	62

CONTENTS

SECT.		PAGE
34.	Function of cyanides	65
35.	White and grey pig	67
36.	Production of malleable iron from pig	69
37.	Bessemer's process	71
38.	Refinery; puddling	73
39.	Rotary puddler	77
40.	Steel	79
41.	Hardening and tempering of steel	80
42.	COPPER: Swansea process	82
43.	Henderson's process	86
44.	Mansfield process, &c.	89
45.	LEAD: Extraction from galena	90
46.	Scotch hearth; refining	90
47.	TIN	92
48.	ZINC: General character and treatment of ores	95
49.	Distillation	96
50.	Separation of cadmium	98

CHAPTER IV.

METALLURGY OF THE LESS IMPORTANT OXIDIZABLE METALS.

51.	ALUMINIUM	100
52.	Preparation from Bauxite	101
53.	MAGNESIUM	103
54.	Use as illuminating agent	104
55.	NICKEL	105
56.	Action of carbon oxide on iron, nickel, and cobalt	107
57.	ANTIMONY	107
58.	BISMUTH	109

CONTENTS.

SECT.		PAGE
59.	ARSENIC	110
60.	MANGANESE	113
61.	Various other metals	114

CHAPTER V.

PHYSICAL PROPERTIES OF METALS.

62.	Lustre; burnishing	115
63.	Manufacture of mirrors: mercurial process for plates	117
64.	Manufacture of mirrors: silver process	118
65.	Formation of double image by glass mirror: Pepper's ghost	119
66.	Colour of metals: by reflection	122
67.	Colour of metals: by transmission	123
68.	Density	123
69.	**Density** of alloys	125
70.	Crystallizability	126
71.	**Malleability** and brittleness	**126**
72.	Goldbeating	**128**
73.	Ductility: wire-drawing	130
74.	Wollaston's method for preparing extremely fine wire	133
75.	Tenacity	134
76.	Influence of physical state, &c., on tenacity	137
77.	Influence of temperature	138
78.	Influence of alloying	139
79.	Other physical properties of metals	140
80.	Pen and pin manufacture	141
81.	Spinning process for making tea-pots, &c.	143

CONTENTS.

CHAPTER VI.

THERMIC AND ELECTRIC RELATIONS OF METALS.

SECT.	PAGE
82. Conductibility for heat	148
83. Conductibility for electricity	149
84. Siemens' pyrometer	150
85. Specific heat : Dulong and Petit's law	152
86. Expansibility	155
87. Exertion of force during contraction	156
88. Fusibility	157
89. Separation of constituents of alloys on standing	159
90. Foundry operations : Bell-founding	161
91. Expansion during solidification	164
92. Annealing, hardening, and tempering	165
93. Volatility	166
94. Thermo-electricity	166
95. Peltier's experiment ; thermo-batteries	168
96. Galvanism or Voltaic electricity	170
97. Magnetism, electro-magnetism, and magneto-electricity	171
98. Bell's articulating telephone	172

CHAPTER VII.

CHEMICAL RELATIONS OF METALS.

99. Special applications of certain metals : platinum	175
100. Protective coatings of less oxidizable metals electro-metallurgy	176
101. Water-gilding, nickel silver, pyro-silver	177
102. Alloys : general features	178

b

CONTENTS.

SECT.		PAGE
103.	Average composition of more important alloys	179
104.	Preparation of alloys : coinage	181
105.	Assaying	182
106.	Poisonous action of copper, lead, tin, &c.	183
107.	Compounds of metals with non-metals	185
108.	Pigments	186
109.	White lead : different processes of manufacture	187
110.	Substitutes for white lead ; zinc white, tungsten white	189
111.	Lakes, mordants, &c.	190

LIST OF ILLUSTRATIONS.

FIG.		PAGE
1.	Cradle for gold-washing	25
2.	Cupel	32
3.	Vertical section of cupel	33
4.	Plan of cupelling furnace	33
5.	Vertical section of ditto	34
6.	Deville's oxyhydrogen furnace	47
7.	Perforation of sheet-iron by oxyhydrogen flame	48
8.	Aludels	50
9.	Palatinate gallery	51
10.	Catalan forge	57
11.	Siemens' gas-producer	59
12.	Blast furnace	60
13.	Bessemer converter	71
14.	Bessemer converter	72
15.	Puddling furnace	74
16.	Dank's rotary puddler	78
17.	Furnace for working poor copper ores by wet process	86
18.	Zinc distillation	97
19.	Oxland's calciner	111
20.	Formation of two images by ordinary glass mirrors	120
21.	Pepper's ghost	121
22.	Graphical representation of different densities of metals	124

LIST OF ILLUSTRATIONS.

FIG.		PAGE
23.	Section of draw-plate	131
24.	Draw-bench	132
25.	Arrangement for illustrating breaking strains of different wires	136
26.	Different stages in pen-manufacture	142
27.	Tea-pot spinning, first stage	144
28.	Sections of chuck and bowled plate	145
29.	Tea-pot spinning, final stage	146
30.	Graphical representation of different expansibilities of metals	154
31.	Bell-founding; preparation of the core	162
32.	Section of moulds and recently cast bell *in situ*	163
33.	Section of Bell's telephone	173

METALS

AND

THEIR CHIEF INDUSTRIAL APPLICATIONS.

CHAPTER I.

METALS AND THEIR NATURAL SOURCES.

1. By various chemical processes the numerous animal, vegetable, and mineral products found in nature can be split up into certain definite constituents, which have hitherto resisted all attempts to decompose them further; to these bodies the term *elements* is applied, substances, into the composition of which two or more elements enter, being designated as *compounds*. Thus by strongly heating a fragment of marble or chalk, the mineral is split up into *quicklime* and a heavy irrespirable gas known as *carbon dioxide;* by special chemical means the quicklime can be shown to be compounded of a brilliant substance termed *calcium*, resembling silver or tin, but rusting much more readily in the air, together with a gas, *oxygen*, lighter than carbon dioxide, and capable of being breathed and of supporting the combustion of inflammable substances with great ease; whilst the carbon dioxide can be similarly split up into charcoal or lamp-

black (chemically designated *carbon*), and oxygen gas identical with that derived from the quicklime. The three substances, calcium, carbon, and oxygen having hitherto resisted all attempts to demonstrate their compound nature, and to break them up into simpler constituents, are therefore termed elements. In similar fashion other elements are obtainable from other natural products; so that finally all the varied products of nature are found to be composed of some or other of about sixty-six elements united together in various proportions.

Of these different elements fourteen differ notably from the remainder in general appearance and texture, and especially in chemical properties : thus hydrogen, oxygen, nitrogen, chlorine, and probably fluorine, are gases at the ordinary temperature, whilst bromine is vaporous at temperatures but little elevated; phosphorus, iodine, sulphur, and its rarer congeners, selenion and tellurion, with carbon, silicon, and boron, though solid at the ordinary temperature, yet differ, in many respects, widely from the other elements, which as a whole possess certain characteristics in common : these fourteen elements are therefore distinguished as *non-metals* or *metalloids*,[1] the others being termed *metals*. Some of the non-metals, and many of the metals, occur on this globe, so far as we are acquainted with it, in much less quantity than others, and are accordingly of comparatively little importance from an industrial point of view on account of their rarity and consequent high price; others, on account of the difficulty of extracting the metals themselves from their natural sources or of special

---

[1] This term is apparently applied on the principle of *lucus a non lucendo*, since the metalloids do *not* resemble the metals save in being elements.

1.] CHIEF INDUSTRIAL APPLICATIONS. 3

properties of certain of their compounds and derivatives, are of less importance as metals than as sources of metallic derivatives of certain kinds.

2. The following table gives the names of those elements the existence of which is well authenticated; it is probable that a few other names should also be added, metals having been described as of rare occurrence, designated *ilmenium, neptunium, lavœsium, &c.;* concerning certain of these confirmatory evidence is, however, wanting. The names printed in capitals are those of metals; those in ordinary type non-metals; the italics in the case of the latter, and the small-capitals in the case of metals, indicate that neither the elements themselves nor their compounds are of great industrial importance; the asterisks denote a greater degree of rarity.

ALUMINIUM.
ANTIMONY.
ARSENIC.
Barium.
BISMUTH.
Boron.
Bromine.
Cadmium.
Calcium.
Carbon.
*Cæsium.
*Cerium.
Chlorine.
CHROMIUM
COBALT.
COPPER.
*Didymium.
*Erbium.
Fluorine.
*Gallium.
*Glucinum.
GOLD.
Hydrogen.
Iodine.
*Indium.

*Iridium.
IRON.
*Lanthanum.
LEAD.
Lithium.
MAGNESIUM.
MANGANESE.
MERCURY.
*Molybdenum.
NICKEL.
*Niobium.
Nitrogen.
*Norium.
*Osmium.
Oxygen.
*Palladium.
Phosphorus.
PLATINUM.
POTASSIUM.
*Rhodium.
*Rubidium.
*Ruthenium.
**Selenion.*
Silicon.
SILVER.

B 2

SODIUM.	TIN.
STRONTIUM.	TITANIUM.
Sulphur.	TUNGSTEN.
*TANTALUM.	*URANIUM.
*Tellurion.	*VANADIUM.
*TERBIUM.	*YTTRIUM.
*THALLIUM.	ZINC.
*THORINUM.	*ZIRCONIUM.

3. From the above list it appears that only twenty-one metals are of any considerable industrial importance, viz., aluminium, antimony, arsenic, bismuth, calcium, chromium, cobalt, copper, gold, iron, lead, magnesium, manganese, mercury, nickel, platinum, potassium, silver, sodium, tin, and zinc. Of the other metals a few are occasionally employed either in the free state, or as compounds of various kinds, for some specific purposes in the arts; thus vanadium compounds have been recently introduced into the calico-printing trade; barium sulphate is largely used as a pigment and especially as an adulterant for whitelead, &c., whilst barium and strontium salts are used in the manufacture of coloured fires for the theatres; certain tungsten compounds have been employed for rendering cotton and other goods uninflammable, whilst others form pigments which will probably be hereafter more largely employed than at present. Palladium is employed to some extent by dentists; cadmium sulphide forms a valuable yellow pigment; uranium compounds give a peculiar colour when added to ordinary glass; and titanium and tungsten form alloys with other metals which may probably be hereafter of considerable commercial importance; but as a rule the industrial value of the other metals and their derivatives is but very small as compared with that of the twenty-one above named.

Of these twenty-one, calcium, chromium, cobalt, and potassium are practically never employed industrially in the metallic or *reguline* state, although their compounds are more or less largely used; the same remark applies to a less extent to sodium; although this metal is somewhat largely manufactured for use in the production of aluminium and magnesium, its compounds are of far more practical importance than the metal itself. Antimony, bismuth, manganese, and, to a lesser extent, arsenic, are used in the metallic state in the form of alloys, whilst certain of their compounds are also of considerable industrial importance; nickel is chiefly employed to prepare certain alloys, especially "German silver," its other compounds being of little practical use; magnesium, when employed in the form of metal, is mainly used as an illuminating agent, its compounds being of much more importance industrially.

4. On the whole, then, the number of metals used to any great extent in the arts in the free metallic state is but limited, being chiefly aluminium, copper, gold, iron, lead, mercury, platinum, silver, tin, and zinc; *i.e.* the seven metals formerly associated with the so-called seven planets,[1] with the addition of aluminium, platinum, and zinc. To these may be further added the metals antimony, bismuth, manganese, and nickel, used almost wholly in the form of alloys. These metals possess to a very high extent the peculiar properties characteristic of the metallic class; most of these properties are also possessed by the less used and the rare metals, and by alloys generally; but in some instances certain of these characteristics are found to be wanting, although the majority of the properties peculiar to metals are possessed by the particular metals or alloys in question:

[1] Sun, gold; Moon, silver; Mercury, mercury; Venus, copper; Mars, iron; Jupiter, tin; Saturn, lead.

these characteristic qualities may be thus summarized: metals possess the power of acquiring under certain conditions a peculiar lustre; they are solid at natural temperatures (mercury freezes in the Arctic regions) and possess (with a few exceptions) at some special temperatures a peculiar coherence and plasticity which enables them to be fashioned into sheets, wires, or other forms, to which property most of their useful applications are due: further they possess certain chemical functions and properties summed up in the term *electro-positive*.

One noteworthy distinction between metals and non-metals is the following,—that when metals are made to mix or unite together, the resulting alloys are frequently formed without any noticeable evolution of heat and invariably still possess the metallic characteristics. Thus, if to melted lead some tin be added, the two metals simply mix together without any indications of chemical action, forming a perfectly homogeneous alloy (solder), just as metallic in its characters as either of the two constituents. If sodium be added to mercury, the evolution of light and heat attests the chemical combination of the two, but the resulting *amalgam* (as alloys containing mercury are termed) is just as much possessed of the characteristic properties of metals as either of its constituents. On the other hand, when a metal unites with a non-metal there is always a greater or lesser evolution of heat, and, with very few exceptions, the product of the combination is destitute of the metallic properties. Thus on burning zinc or iron in oxygen gas, or magnesium in the air; on throwing antimony filings into chlorine gas;[1] or on warming a mixture of copper or iron filings

---

[1] It is peculiarly noteworthy that whilst combinations between metals and non-metals occur with great force when once started, a

and sulphur, a rapid combination between the metal and non-metal employed is brought about with the evolution of much heat and light, and products are formed in each case respectively, wholly destitute of metallic characters.

In some few instances, however, small quantities of certain non-metals can be incorporated with various metals and alloys without destroying their peculiar characteristics; in the case of steel and Bessemer metal, the presence of a small quantity of carbon in the iron or iron-manganese alloy even augments the strength and tenacity; and the same is true as regards phosphorus in phosphor-bronze.

Some few of the more important metals are occasionally found in nature in the free state, either alone or alloyed together, and not combined with any non-metallic element: such metals are said to be *native*. Ordinarily, however, metals are met with associated with non-metallic elements in the form of compounds, which are termed *ores*; the extraction of metals from their ores constitutes the chemical art of *metallurgy*, the general principle of which is the reversal of the phenomena which occur when metals and non-metals unite together, either by direct or indirect processes; thus, when combination occurs the phenomenon may be typified by the symbols

$$A + B = AB,[1]$$

certain temperature is essential in order to cause the action to commence. Thus, pure dry oxygen has no action on iron or zinc at the ordinary temperature; nor does sulphur act on copper until the mixture is gently heated; whilst at a temperature of $-80°$, liquid chlorine has no action on antimony.

[1] Juxtaposition of symbols, as AB, means that the constituents A and B are chemically combined together; whilst the separation of the symbols by the sign + indicates that these constituents are separate from one another.

the mode of action being the kind termed *synthetic;* the processes of metal extraction on the other hand are either of the kind spoken of as *analytic,* typified by the symbols

$$AB = A + B$$

(as, for example, when oxide of silver breaks up on heating into oxygen and silver, or when copper chloride is electrolysed into copper and chlorine); or belong to the class of chemical actions known as *reactions of single decomposition,* represented by the symbols

$$AB + C = A + BC$$

(as, for example, where hydrogen is passed over heated silver chloride, forming silver and hydrogen chloride). This kind of reaction is by far the most common in practical metallurgy.

A fourth kind of chemical action, known as *double decomposition,* and represented by the symbols

$$AB + CD = AC + BD,$$

is occasionally an intermediate step in the isolation of a metal from an ore, the object being to change one kind of metallic derivative into another more easily treated (for example, where silver sulphide is transformed into silver chloride by the agency of copper chloride as a preliminary stage in the amalgamation process, as applied to ores containing silver as sulphide). This kind of action is, however, more frequently utilized in the manufacture of metallic derivatives, and notably of various pigments formed by precipitation (§ 108).

6. Apart from the nature of the chemical changes involved, metal-extracting processes may be divided into the following classes, according to the physical properties

of the metals themselves and the general character of the process.

Volatilization Processes: applicable when the metal is capable of being converted into vapour and recondensed at temperatures which can be readily obtained in practice: such metals are *zinc, mercury, sodium, potassium, arsenic, magnesium, cadmium*, and some others. Silver can be distilled in a lime crucible by the oxyhydrogen flame (Stas), and lead is sensibly volatilized during the smelting of galena, the volatilized metal being condensed again as oxide, &c., constituting the "fume" of the lead smelter. Other metals, *e.g.*, copper and gold, are sensibly volatile at high temperatures, but this property is so faintly marked that it is not capable of being utilized in their treatment.

Clearly, where it is necessary to separate two metals, one of which is volatile and one not, all that is requisite is to heat the alloy sufficiently to expel the volatile metal, which can be either recondensed or not as convenient. In this way platinum-arsenic alloys have been proposed for the manufacture of platinum articles, the alloy being more fusible than pure platinum; by heating the manufactured article (*e.g.*, a crucible) the arsenic is gradually expelled; similarly, from a ternary alloy of silver lead and zinc the zinc can be removed by distillation (*vide* § 20).

Amalgamation Processes.—These are really special cases of the volatilization processes; the ore is treated with some chemical agent that will convert it into the metallic state (if not already in that condition) and then brought into contact with mercury, which dissolves the metal: from the resulting amalgam the mercury is separated by distillation. On account of the loss of mercury, and the cost of this body, this kind of process can only be economically applied to the precious or the rarer metals.

Smelting Processes.—When the metal is fusible at a moderate or high temperature, it is frequently extracted from its ores by the application of some chemical agents which will, at some more or less elevated temperature, act on the metallic compound employed, causing the metal to become free in the liquid form. *Copper, iron, lead, antimony, aluminium,* and *tin,* in particular, are usually obtained by processes of this kind.

Liquation Processes: modifications of the preceding, used to separate a readily fusible metal in the free state either from an infusible rocky matrix or *gangue* (*e.g.* bismuth), or from another less fusible metal (*e.g.* copper and lead). The mass is simply cautiously heated, when the more fusible metal gradually melts and runs away from the other less fusible substance. In the purification of crude tin as obtained by the smelter, this process is largely employed, the tin melting first and running away from the other metals present.

Wet Processes: where the metal is extracted by some method which necessitates the application of chemical reagents to an aqueous solution of a compound of the metal. In the treatment of complex ores this mode of operating is often combined with some process or processes belonging to other kinds.

Miscellaneous Processes, *e.g.* where gases or other reagents are made to react on a metallic compound so as to yield the metal in the free state without fusion: thus, in the preparation of *ferrum redactum* (iron reduced by hydrogen): *nickel; silver* from the chloride by contact with zinc, &c. Or where *platinum* is obtained from ammonium platinochloride, or *silver* from silver oxide, by simply heating the compound: and so on.

I.]   CHIEF INDUSTRIAL APPLICATIONS.   11

7. The following table illustrates the general chemical characters of the chief processes employed for the purpose of the isolation from their natural sources of metals of ordinary occurrence; before, however, these processes can be applied, the natural minerals have frequently to be subjected to some preliminary mechanical or chemical treatment : thus, most ores, as well as gold quartz (native gold disseminated in minute grains throughout a large mass of quartzose rock), &c., are crushed or stamped ; the lighter earthy particles are often separated by washing with water or otherwise; the crude ores are frequently calcined for the purpose of either producing some chemical change in the material (*e.g.* clay ironstone, where the ferrous carbonate becomes converted into ferric oxide by calcination), accompanied by a physical alteration in texture so as to render the ore more tractable in the subsequent stages ; or of removing small quantities of adherent or admixed ores of other kinds (as with tinstone, which is calcined and then washed for the purpose of converting any iron pyrites, &c., originally present into ferric oxide by calcination, and removal by washing of the lighter oxides so produced).

General Character of Metal-extracting Processes in Common Use.

A.—NATIVE METALS.

By mechanical means   ..   .. *e.g.* gold-washing.
,, simple fusion (liquation)   ..   ,, bismuth.
,, solution in mercury   ..   .. ,, gold-quartz.
,, solution in aqueous chemicals   ,, ditto.

B.—SIMPLE ORES ; *i.e.* containing only one metal.

I. OXIDES.

Analytic .. .. By simple heating   ..   ..   .. *e.g.* mercury, silver.
Single decom- ⎧ ,, heating in hydrogen   ..   ..   ,, nickel, iron.
position   .. ⎨ ,, heating in carbon oxide   ..   ,, iron (blast-furnace).
               ⎩ ,, heating with carbon (coal⎫ ⎰tin, arsenic, zinc, iron,
                     coke, &c.) .. ..⎭ '' ⎱ antimony.

II. CHLORIDES, FLUORIDES, &c.

Analytic	By heating alone	e.g. platinum, gold.
	,, heating in hydrogen	,, silver.
Single decom- position ..	,, action of cheaper metal, &c.	
	,, (a) wet processes	,, copper, gold.
	,, (b) dry processes	,, magnesium, aluminium.
	,, (c) amalgamation processes .	,, silver.

III. SULPHIDES.

Single decom- position ..	By heating with air	e.g. mercury, copper, lead.
	,, heating with cheaper metal..	,, mercury, antimony, lead.
Double decom- position fol- lowed by sin- gle decompo- sition	,, roasting to oxide and reduc- ing as above	,, iron, zinc, antimony.
	,, converting into chloride and treating as above	,, silver.

IV. CARBONATES.

Single decom- position.. ..	By heating with carbon	e.g. zinc, sodium, potassium.
Double decom- position fol- lowed by sin- gle decompo- sition	,, roasting to oxide and reduc- ing as above	,, iron.
	,, converting into chloride and treating as above	,, copper

C.—COMPLEX ORES; *i.e.*, containing more than one metal.

I. Alloy extracted by some or other process, as above ..	e.g.	silver-lead alloy, spie- geleisen.
II. Special processes adopted for extraction of metals sepa- rately	,,	cupriferous pyrites.

8. Native Metals.—Bismuth, copper, gold, iron (associated with nickel and other metals in small quantities), mercury, platinum (and allied rarer metals), and silver are the metals most frequently found in the native state. Iron, when native, is usually of extra-terrestrial origin, *i.e.* it occurs as *meteorites;* copper appears to owe its occurrence in the native state to galvanic action (thermo-electric currents?) whereby soluble copper compounds formed naturally have been decomposed and metallic copper separated precisely as in ordinary electrotyping; and the same kind of action is probably the cause of the occurrence of other native metals. The metals which most frequently occur native are for the most part, however, members of the class termed *precious* or *noble metals*, from their feeble affinity for oxygen, which renders them incapable of rusting or tarnishing by oxidation in the air at common temperatures; thus neither at the ordinary temperature, nor on heating, will platinum, gold, or silver rust or oxidize (or combine with oxygen); on the contrary, the oxides of these metals cease to exist on heating, the metal being set free, and oxygen being given off as a gas. It is noticeable, however, that the inability to combine directly with oxygen, and the instability of the oxide producible by indirect chemical means, exhibited by the noble metals, does not necessarily extend to their combinations with other non-metals: thus, whilst gold chloride readily breaks up on heating into gold and chlorine, silver chloride can be fused unchanged at a red heat, and mercuric chloride can be volatilized and recondensed without decomposition; whilst gold and chlorine will readily unite at the ordinary temperature. Similarly, though sulphide of ammonium does not attack platinum, it blackens silver (by yielding sulphur thereto) just as readily as copper at the ordinary temperature; and to this ready sulphuration of silver is due the blackening of silver goods on lying by for some time.

9. **Simple Ores.**—The first step towards isolating a metal from any particular ore consists in obtaining this ore in a reasonably pure state, *i.e.* in separating from the ore itself foreign particles of the earthy or rocky matter in which it is naturally embedded, technically termed the *matrix*, or *gangue*. Frequently this is effected by hand; the lumps of material raised from the mine being dressed by a hammer, and the fragments picked over so as to separate moderately completely the gangue: in some cases the ore itself, being much heavier than the gangue, is separated therefrom by stamping to a coarse powder, and then subjecting this to a process of washing analogous to that of gold washing (§ 15); special machinery is usually

employed for this purpose. When the ore has been rendered tolerably free from earthy admixtures, &c., by washing or otherwise (and in some cases after a preliminary calcination or other chemical treatment as mentioned above), it is subjected to the action of some suitable reagent which will take away the associated non-metallic constituents by a chemical process, the end-result [1] of which is represented by the equation of single decomposition.

$$AB + C = A + BC.$$

[1] Although the end-result of the reaction of a substance C on a compound body AB may be such as to be represented by the above equation, yet it often happens that the result is only brought about *as the final sum of a series of chemical changes:* in operating on the manufacturing scale, with tons of material in large and frequently enormous vessels, it often happens that these successive changes may be traced as representing clearly-defined separate stages in the process. For example, in the extraction of lead from galena, the end-result is that represented by the equation—

(1) $PbS + O_2 = Pb + SO_2$,

i.e. where lead sulphide and oxygen yield metallic lead and sulphur dioxide; but this final change is brought about as the sum of several intermediate stages. A portion of the lead sulphide becomes converted into lead sulphate, thus—

(2) $PbS + 2O_2 = PbSO_4$,

and this sulphate reacts on the undecomposed sulphide, forming sulphur dioxide and metallic lead—

(3) $PbSO_4 + PbS = 2SO_2 + 2Pb$.

Equation (1) is thus equivalent to the sum of equations (2) and (3). Another portion of the lead sulphide becomes converted into lead oxide, thus—

(4) $2PbS + 3O_2 = 2SO_2 + 2PbO$,

and this oxide reacts on the **unaltered** galena, again forming sulphur dioxide and metallic lead—

(5) $2PbO + PbS = SO_2 + 3Pb$.

Here, again, equations (4) and (5) jointly are equivalent to equation (1). In the smelting of iron oxide by the blast-furnace the intermediate stages are yet more numerous (*vide* § 33).

I.]   CHIEF INDUSTRIAL APPLICATIONS.   15

In this way are obtained the metals nickel, iron, tin, arsenic, silver, copper, gold, magnesium, aluminium, mercury, lead, antimony, zinc, sodium, potassium, &c., &c.

10. **Complex Ores.**—In the majority of the processes to which these substances are subjected, the actions are much less simple than the foregoing cases of simple ores. Complicated series of operations have frequently to be gone through for the purpose of separating one metal or metallic compound from another; thus in the extraction of copper from copper pyrites by the "dry process," the iron contained in the pyrites is separated by a series of operations, the end-result of which is to oxidize the iron and part of the sulphur contained in the mineral, and to form a compound of copper and sulphur, from which the copper is subsequently extracted by a further series of operations. In one of the "wet processes," for treating pyrites containing copper and silver, the pyrites is first roasted to expel the majority of the sulphur and to oxidize the iron present; the residue is then heated with common salt, whereby the copper is transformed into chloride (as are also the silver and a small portion of the iron). On treating with water or diluted hydrochloric acid the product of the action, the copper, silver, and iron chlorides formed are dissolved out, the silver chloride (if, as is usually the case, it is present in only small quantity) being retained in solution by the portion of common salt which has remained unacted upon; from the mixture of solutions of various chlorides thus obtained, the metals are separated by further chemical processes (*vide* §§ 21 and 43), whilst the oxide of iron left undissolved is employed as a source of metallic iron, or for other special purposes.

In some instances complex ores are treated by processes the end-result of which is to yield an alloy of two or more metals which are (if required) then separated by further operations: thus, spiegeleisen and ferro-manganese (alloys of iron and manganese with additional carbon), largely used in the production of Bessemer metal, are smelted from manganiferous iron ores in just the same way as iron ores themselves (minor modifications of the processes due to physical differences in the ores, &c., excepted). Ores containing copper, lead, and silver are smelted together so as to obtain a ternary alloy of these metals from which a silver and lead alloy is separated by liquation, the silver being subsequently isolated by cupellation of this alloy so obtained (*vide* § 19): or again from a binary alloy of lead and silver, the silver is removed by a somewhat roundabout process, viz. mixing with fused zinc, which dissolves out, as it were, the silver from the lead and floats up to the top, much as ether when shaken up with water containing various substances in solution, will float up to the top after dissolving these bodies and separating them from the water: from the silver-zinc alloy thus produced (also containing a little lead) the zinc is thus removed by distillation, and the residual silver and lead separated by cupellation (*vide* § 20).

11. When the relationships are studied that exist between the circumstances under which a chemical change of the kind termed single decomposition can take place, and the proportionate disturbance of the thermal equilibrium prought about during such reactions, some curious correlations are noticeable. If a body A, in uniting with a given weight of a body B, evolve a certain amount of heat H_1, and a body C in uniting with the same weight of B, evolves an amount of heat H_2, the thermal disturbance during the reaction

$$AB + C = A + BC$$

will clearly be an evolution of heat, the value of which is $H_2 - H_1$, inasmuch as in undoing the combination AB, H_1 of heat must be absorbed; whilst in producing the combination BC, H_2 of heat is evolved. It may happen, however, that H_1 is greater than H_2, so that the value of $H_2 - H_1$ is *negative*; that is to say, instead of an evolution of heat during the reaction the chemical change is accompanied on the whole by an absorption of heat.

When the body A in the above general equation is a metal, and the reaction takes place in an aqueous solution (*i.e.* a *wet* process, and hence not at an elevated temperature), the value of $H_2 - H_1$ *is never negative*, C being another metal, if indeed (which is doubtful) it ever be negative whatever C be. In other words, *one metal can only displace another one from a solution of one of its compounds when the second metal evolves more heat in uniting with the other constituents of the compound than does the first metal.* Thus, in the familiar experiments of the lead-tree and the *Arbor Dianæ*, or silver-tree (where lead is displaced from lead acetate by zinc, forming zinc acetate, and silver from silver nitrate by mercury, forming mercury nitrate), the chemical changes can take place because they are on the whole accompanied by evolution of heat. Similarly from the mercury nitrate solution obtained in the *Arbor Dianæ* copper will reprecipitate the mercury in solution, forming copper nitrate, from which, in turn, the copper may be thrown down by zinc or by iron, each successive change being accompanied by an evolution of heat.

12. In reactions of the above general character, when the substances employed are not in solution, the reduction of a metal is sometimes accompanied by a marked heat absorption, although the reverse is more frequently the case. Whenever any considerable heat absorption

occurs it is noticeable that the reduction *never takes place at low temperatures;* the reaction is only brought about when the substances are heated to some considerable extent, *the temperature of commencing action being usually higher the greater the heat absorption.* Similarly the temperature at which action commences in cases where there is heat evolution *is usually the lower the greater the value of the heat evolution.* The character of the molecular condition of the metallic compound, however (whether light and porous, or dense and heavy), considerably affects the temperature of initial action.

Thus, for example, the action of carbon oxide on metallic oxides, such as copper, iron, and zinc, whereby carbon dioxide and the metals are formed, is attended with very considerable differences in heat evolution in each case: the "heats of combustion" of these metals and of carbon oxide are as follows:—

Metal or other combustible.	Kilogramme heat units evolved by the union of 16 grammes of oxygen with the combustible.	Oxide formed.	Authority.
Copper	38·30	Cupric oxide, CuO	Andrews.
Iron	66·45	Ferrosoferric oxide, Fe_3O_4	,,
Zinc	86·24	Zinc oxide, ZnO	,,
Carbon oxide	68·35	Carbon dioxide, CO_2	Mean of results of Andrews, Favre and Silbermann, and Dulong.

Whence it results that if the reactions:

$$CuO + CO = Cu + CO_2$$
$$Fe_3O_4 + 4CO = 3Fe + 4CO_2$$
$$ZnO + CO = Zn + CO_2$$

could take place at the ordinary temperature, the heat disturbances (evolution or absorption of heat) accompanying them would be respectively (per 16 grammes of oxygen transferred from the metallic oxide to the reducing agent):—

Kilogramme
heat units.

Reduction of copper oxide ... $68\cdot35 - 38\cdot30 = + 30\cdot05$
,, iron ,, ... $68\cdot35 - 66\cdot45 = + 1\cdot90$
,, zinc ,, ... $68\cdot35 - 86\cdot24 = - 17\cdot89$

That is, the reduction of the copper oxide would be attended with a large heat evolution, that of the iron oxide with a small heat evolution, and that of the zinc oxide with a considerable heat absorption: but little difference in the relative values ensues if the heat disturbances be calculated on the supposition that they take place at somewhat more elevated temperatures, 200° or 300° C. for example.[1] Now it has been recently

---

[1] These calculated values are easily obtained by means of the formula,
$$H_\tau = H + (h_1 + h_2) - (h_3 + h_4),$$
where H_τ is the value of the heat disturbance when the reaction takes place at T° (*i.e.* all the materials and products being at the temperature T), H being the heat disturbance at the ordinary temperature (15° or thereabouts).

h_1 is the heat required to raise the metallic oxide from 15° to °T
h_2 ,, ,, ,, carbon oxide ,, ,, ,,
h_3 ,, ,, ,, { carbon dioxide produced } ,, ,, ,,
h_4 ,, ,, ,, metal set free ,, ,, ,,

The values of h_1, h_2, h_3, h_4 are calculated by the formula—
$$h = W \times S \times (T - 15),$$
where W is the weight of the substance supposed to be heated, and S its specific heat between 15° and T°.

Usually the correction $h_1 + h_2$, itself not very large, is almost balanced by the similar correction of opposite sign $h_3 + h_4$.

shown by Alder-Wright and Luff[1] that the action of carbon oxide on copper oxide begins at a lower temperature than that on ferric oxide, when the molecular conditions of the two oxides are comparable; whilst both these kinds of oxides are affected by carbon oxide at much lower temperatures than zinc oxide. Thus Lowthian Bell found[2] that carbon oxide does not reduce zinc oxide at all at 420°; whilst Alder-Wright and Luff obtained the following numbers as the temperatures of initial action of carbon oxide on copper and iron oxides :—

Oxides prepared by precipitation.

	Copper oxide.	Iron oxide.
Temperature at which action commences	60°	90°

Oxides prepared by the ignition of salts.

	Copper oxide.	Iron oxide.
Temperature at which action commences	125°	$\begin{cases} 202° \\ 220° \end{cases}$

Precisely similar results are obtained when other reducing agents are substituted for carbon oxide; the absolute values of the heat disturbances are, of course, different for each reducing agent, but with a given reducing agent the same relationship necessarily holds as with carbon oxide, viz., that the value of the heat evolution is algebraically greater with copper oxide than with iron oxide. Thus the state of aggregation of the metallic oxides being the same, or nearly so, the following values were obtained as the temperatures of initial action of hydrogen and carbon :—

[1] *Journal of the Chemical Society,* January, 1878.
[2] *Chemical Phenomena of Iron Smelting,* p. 91.

Reducing agent.	Character of oxide used.	Temperature of initial action.	
		Copper oxide.	Iron oxide.
Hydrogen	Precipitated	85°	195°
Ditto	By ignition of salts	175	260
Carbon from sugar ...	Precipitated	390	450
Carbon from another source (much lighter)	By ignition of salts	390	430
Ditto ditto ...	Precipitated	350	430

In all cases the temperature of initial action on copper oxide is sensibly lower than that on iron oxide in an analogous state of aggregation.

13. Again, on comparing the amounts of heat developed during the reduction of a given metallic oxide in a given molecular state by various reagents, it is noticeable (as might indeed be expected from the preceding results) that *the reducing agent which causes a heat disturbance of higher algebraic value uniformly begins to act at the lower temperature;* i.e. if there be heat evolution, the greater the evolution the lower the temperature of initial action; and if there be heat absorption, the less the absorption the lower the temperature of initial action. Thus, the heat disturbances during the reduction of copper and iron oxides by carbon oxide, hydrogen (steam being formed), and free carbon (carbon dioxide being produced) are respectively :—

	Copper oxide.	Iron oxide.
Carbon oxide . .	+ 30·05	+ 1·90
Hydrogen . . .	+ 19·52	− 8·63
Carbon	+ 9·48	− 18·67

since the "heats of combustion" of hydrogen and carbon (to steam and carbon dioxide respectively) are 57·82 and 47·78 kilogramme heat units per 16 grammes of oxygen added on to the combustible;[1] accordingly Alder-Wright and Luff found the following values for the temperatures of initial action of these reducing agents on several sorts of copper and iron oxides prepared so as to present considerable variation in their state of molecular aggregation.

OXIDES OF COPPER.

Reducing agent.	Precipitated Cupric oxide.	From nitrate by calcination.	By continued roasting of metal.	Cuprous oxide.
Carbon oxide ...	60	125	146	110
Hydrogen	85	175	172	155
Carbon (light) ...	350	390	430	345
,, (dense)...	390	430	440	380

OXIDES OF IRON.

Reducing agent.	Precipitated Ferric oxide.	Precipitated and gently ign.ted.	From ignition of Sulphate.
Carbon oxide ...	90	220	202
Hydrogen	195	245	260
Carbon (light) ...	—	430	430
,, (dense)...	450	450	450

The action of carbon oxide uniformly beginning at a lower temperature than that of hydrogen, which again commences at a temperature lower than that requisite in

[1] Mean results of a large number of determinations by several chemists (vide Alder-Wright, *Phil. Mag.* Dec., 1874). In the case of hydrogen, the average heat of combustion to liquid water per 16 grammes of oxygen = 2 × 34·275 is reduced by 18 × 0·596 = 10·73, the latent heat of aqueous vapour at 15° being 596 (Regnault).

the case of carbon. From experiments not yet published it appears that precisely analogous results are given with other metallic oxides.

14. In some cases it is noticeable that if the heat disturbance during the reduction of a metallic oxide is not of considerable magnitude, *the reaction can be inverted by modifying the conditions of the experiment;* i.e. if under certain conditions the reaction,

$$AB + C = A + BC$$

can occur, under other conditions the opposite reaction,

$$A + BC = AB + C$$

can take place; this, however, does not very frequently occur. Two of the best illustrations of the phenomenon are afforded by iron oxide with carbon oxide and hydrogen respectively; the heat disturbances during the reduction of the metallic oxide by these two agents are $+ 1\cdot 90$ and $- 8\cdot 63$ (§ 13); accordingly, if hydrogen or carbon oxide in excess is passed over heated iron oxide reduction of the metallic oxide and oxidation of the reducing agent will take place; but if excess of carbon dioxide or of steam be passed over heated metallic iron, oxidation of the metal occurs together with formation of carbon oxide in the one case, and of hydrogen in the other.

Nickel and cobalt yield similar results (Lowthian Bell).

CHAPTER II.

METALLURGY OF THE PRECIOUS OR NOBLE METALS.

15. THE processes in actual use for the extraction of metals generally from their natural sources are so numerous and vary so much in details, according to circumstances, that only a brief outline of some of the more important methods in use in the case of the chief metals can be attempted here.

Gold.—For the most part this element is found native (§ 8), and generally in a state of considerable freedom from other metals, the nature of the accompanying metals, when present, varying with the locality. Silver is very frequently associated with native gold; that from Australia and California often containing 5 to 10, or even more, per cent. of this metal, with small quantities of copper and iron in addition. Brazilian gold often contains palladium, and that from Russia platinum. In Hungary gold is found associated with tellurion, whilst the Huelva and Tharsis pyrites and many other minerals contain small quantities of this precious metal interspersed throughout masses of other metallic sulphides. As a rule, the metal occurs in quartzose and granitic rocks, but large quantities have been obtained from time immemorial from the sands of river-beds and from alluvial deposits of the matters washed down by streams from mountainous districts where these rocks occur. Occasionally in these deposits the gold is found in smaller or larger masses known as *nuggets*, some of which have been found of considerable magnitude, weighing upwards of one, and even of two cwt., and worth several thousand pounds sterling; generally, however, the gold occurs in minute grains or "dust." From alluvial soils the gold is separated by simple mechanical means, the earthy mass being agitated

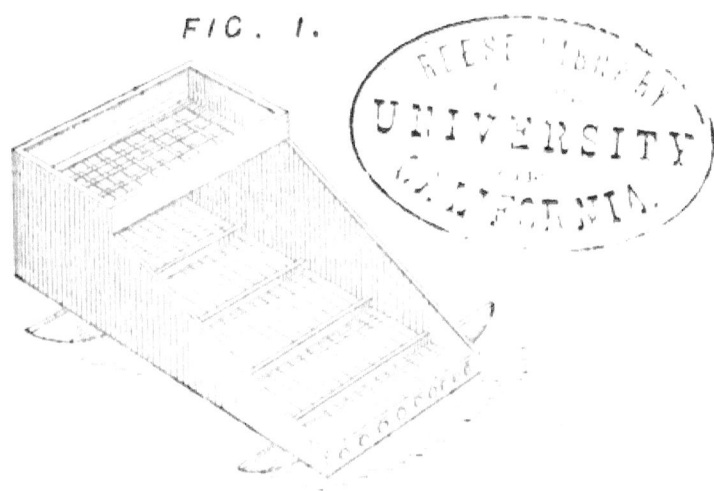

FIG. 1.

with water in a wide shallow basin, technically termed a "pan," a peculiar motion being communicated by the hand so as to scoop out of the pan the muddy water whilst the heavier particles of metal are retained. When necessary, stones and hardened masses are previously roughly pulverized, or they are mechanically picked out by the hand during washing. For systematic working a washing-machine, known as a "cradle," is employed (Fig. 1), consisting of a kind of wooden box, some 6 or 7 feet long, and about 2 feet wide, mounted on rockers so that the bottom has a gentle slope from one end to the other; across the bottom are nailed wooden bars (riffle-bars) so as to make a series of small shallow weirs in the cradle; at the top end is fixed a sieve, on to which the earth is shovelled, a constant stream of water being kept running on to the mass; the larger stones are thus left on the sieve, whilst the clayey portion, the small gravel, and the gold dust are washed through and deposited in the shallow pools formed by the riffle-bars. Every now and then the deposited particles are removed and dried in the sun, and the lighter earthy particles blown away by the breath, and the small pebbles picked out. Where the soil is a stiff clay, the mass is stirred up with water in a tub with paddles or revolving arms before cradling; and, occasionally, instead of carrying the washing so far as to separate all the earthy matters (which often entails loss by the washing away of the finest particles of gold), the cradling is only carried on to a certain point, the gold being extracted from the partially washed mass by amalgamation; this operation, however, is more commonly employed in the case of the powder obtained by stamping or grinding quartzose auriferous rocks, and the washing is then more frequently effected by making the water and suspended matters pass over a series of blankets stretched tightly on frames so as to make a series of very slightly inclined planes; the woollen surfaces arrest the particles of gold with more or less earthy matters. From time to time the deposit is removed, and the gold extracted by means of mercury. Occasionally a kind of magnified cradle is used, consisting of a series of slightly inclined wooden troughs or sluices, 10 or 12 feet long, so that the stream from one is delivered on to the next, and so on. Riffle-bars are placed in these troughs either directly across or at an angle. When the washing has gone on for a short time mercury is allowed to flow into the upper trough; this runs down, being retained by the riffle-bars, so that the mercury dissolves the accumulated particles of gold. In order to retain the finest particles of gold which sometimes escape through being washed down too rapidly to be absorbed by the little pools of mercury in the sluices, amalgamated plates of copper are placed at the end of the lowest trough, so that the effluent "slimes" must pass over them; the amalgam of gold and copper thus formed is from time to time removed and worked up with the fluid mercury amalgam.

16. In the extraction of gold from gold quartz, the stamped rock (washed by blankets or sluices when requisite) is placed in an iron pan with a certain quantity of mercury; a stream of water flows into the pan, the overflow passing into another pan at a lower level also containing mercury. Several pans are generally connected in series, each being provided with an agitator worked by steam or water power after the fashion of a mortar mill; the gold is then dissolved out, the agitator continually bringing the auriferous particles and the mercury in contact. Fresh quantities of stamped quartz are continually introduced until the mercury has taken up so much gold as partially to lose its fluidity; the amalgam is then squeezed through chamois leather, whereby a fluid amalgam, containing only a small quantity of gold, is separated, and a nearly solid rich amalgam retained. The fluid amalgam is used over again in the pans, the solid mass being carefully distilled so as to separate the mercury and leave the gold and other metals dissolved out. Certain auriferous minerals, when treated with mercury, do not allow the gold to be wholly dissolved out,

the gold becoming sulphurized or otherwise changed on the surface so as to prevent contact between the gold and the mercury. This is especially the case with ores containing sulphur, arsenic, or tellurion. To avoid this it has been proposed by Crookes, and also by Wurtz (of New York), to add small quantities of sodium to the mercury. The sodium amalgam thus obtained causes the mercury to wet the metallic particles (probably by destroying the film of gold sulphide or by preventing its formation). An additional advantage in this process is that the loss by "flouring" of the mercury (reduction to fine particles which do not again coalesce and are consequently washed away) is to a large extent prevented; the mercury being always clean and bright, any small particles mechanically dislodged are readily reabsorbed into the rest of the mass of mercury.

17. When obtained from certain ores, the gold thus extracted is apt to contain small quantities of foreign metals which destroy its tenacity and other valuable properties; in many cases these metals, if present in the mass left on distilling off the mercury, can be removed by simply melting the residue in crucibles with an oxidizing flux, such as a mixture of nitre and borax; the foreign impurities are thus removed, being oxidized by the nitre, the oxides thus produced being taken up by the melted borax. Another method of purifying gold, especially suited for gold rendered brittle by the presence of tin and antimony, was introduced a few years ago by Mr. F. B. Miller of the Sydney Mint; this consists of passing a stream of chlorine gas through the molten metal, when silver and baser metals are converted into chlorides, the latter being expelled if volatile, whilst the gold remains unchanged; in this way perfectly pure gold is readily prepared, whilst the silver chloride formed can

be readily reduced to the metallic state and the silver thus obtained separate from the gold. When the silver, in an alloy of gold and silver, largely preponderates, as is the case when silver containing small quantities of gold is extracted from such sources as Spanish pyrites, &c., a more convenient mode of separating the gold is to treat the alloy with nitric or sulphuric acid, when the silver dissolves and the gold is left undissolved (§ 24). In the assaying of coinage alloys this mode of separation is largely used (*vide* § 105).

18. A mode of extracting gold by purely chemical means from gold quartz and other similar sources has been proposed by Calvert : the crushed quartz is subjected to the action of cold chlorine gas so as to convert the gold and other metals present into chlorides which are subsequently washed out by water ; from the solution thus obtained metallic gold in a pure state is separated by adding solution of ferrous sulphate, or better, by blowing sulphur dioxide through the liquid ;[1] the gold thus separates as a fine powder which is allowed to subside, washed, collected, and fused into an ingot. If silver be also present the metal may be extracted by treating the chlorinated mass with brine, which dissolves the silver chloride, and subsequently separates the silver from the solution thus obtained (§ 21). Another process, patented by Longmaid, for separating gold from substances containing it consists in fusing the ore in a reverberatory furnace with roasted pyrites, lime, and fluor spar ; some gold subsides to the bottom, whilst some remains suspended and is separated by placing iron plates in the fused mass ;

---

[1] The chemical changes thus taking place may be written :—

$$2AuCl_3 + 6FeSO_4 = 2Au + Fe_2Cl_6 + 2Fe_2(SO_4)_3,$$
$$2AuCl_3 + 3SO_2 + 6H_2O = 2Au + 6HCl + 3SO_4H_2.$$

the gold adheres to these and is then dissolved off from them by immersing them in a bath of melted lead. In this way there is ultimately obtained a gold-lead alloy, from which the gold is separated by cupellation (§ 19). Neither this process nor Calvert's seems to have come into any considerable practical use, although a process much the same in principle as the latter has been worked in Silesia; this method, described by Plattner in the Jurors' Report of the 1851 Exhibition, consisted in roasting the ore (an arsenical pyrites containing gold to the extent of somewhat less than $\frac{1}{2}$ oz. per ton) so as to drive off arsenic and sulphur, treating the residue with chlorine gas, and then dissolving out by water the chlorides of iron and gold and separating the latter by a reducing agent such as sulphuretted hydrogen.

19. **Silver.**—Besides occurring native in Mexico, Chili, and Peru, and in smaller quantities in many other places, this metal occurs in combination with sulphur, forming several kinds of ores according to the quantity and character of the other associated substances; also as chloride (horn silver), iodide, arsenide, bromide, and as antimony and mercury alloys. Considerable quantities are now extracted from the lead smelted from lead sulphide containing small quantities of silver; whilst the Spanish pyrites largely used for the production of sulphuric acid yields a residue on burning off the sulphur from which silver and copper are obtained in some quantity. From these sources the silver is extracted by methods which naturally fall under one or other of the three classes known as *lead processes, wet methods*, and *amalgamation processes*, the distinctive feature of the first class being that from a suitable mixture of ores a silver-lead alloy is smelted, from which the precious metal is subsequently separated; whilst in the second class the silver is obtained

in aqueous solution by treatment with appropriate reagents and is thence precipitated by chemical agency; and in the third class the silver is extracted by means of mercury somewhat after the fashion of gold (§ 16), but usually by processes involving more complex chemical actions.

Lead processes.—When silver exists in the metallic state disseminated through a rocky matrix, a silver-lead alloy is readily formed by merely fusing together the ore and a sufficient quantity of metallic lead, whereby the silver is, as it were, dissolved out of the matrix by the molten lead, much as gold is out of gold quartz by mercury. When the silver is not present in the metallic state, a silver-lead alloy can be obtained by mixing with the silver ores, galena (lead sulphide), or other lead ores and then smelting the whole together so as to reduce both lead and silver to the metallic state. From the argentiferous lead thus obtained the silver is then extracted by processes the exact character of which depends upon circumstances; if the silver-lead alloy be sufficiently rich in silver and tolerably free from copper, the lead and other foreign base metals present in greater or less quantity are removed by a process termed "cupellation," which consists in fusing the alloy in a shallow dish (*cupel*) made of porous material, such as compressed bone-ash, clay, or marl, moistened with wood-ash liquor, whilst a stream of air plays over its surface; the lead and foreign base metals oxidize and the oxides fuse and are partially absorbed by the cupel and partially blown away by the current of air or drawn off in the liquid state forming the lead oxides commercially known as *litharge* and *massicot*. Finally, when all the foreign metals are oxidized a mass of fused silver is left containing in addition any gold or platinum which may happen to have been originally present; the clearing away of the last traces of oxide produces a rather singular appearance on the surface of fused metal, technically termed the "brightening." The fused silver, especially at high temperatures, absorbs oxygen from the air and on cooling is apt to give it out again, forming peculiar miniature volcanic cones where the escape of gas forces the yet liquid metal in the interior up through the newly-formed thin solid crust; this phenomenon is spoken of as "vegetation," or "spitting," and with large masses of silver (one cwt. or so) is often extremely well defined, the cones being an inch or two in height.

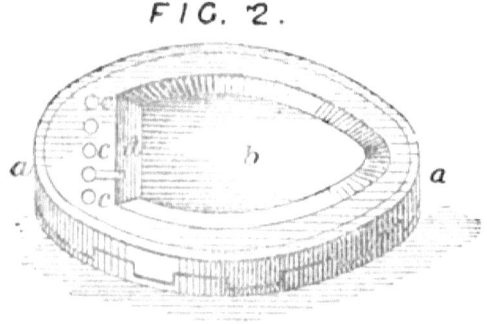

FIG. 2.

Fig. 2 represents the form of cupel usually employed in this country; an egg-shaped ring of half-inch boiler plate, *a, a*, strengthened by cross-bars at the bottom, constitutes a sieve-like frame into which is rammed by mallets a mixture of bone-ashes and a little wood-ash moistened with weak potash solution; the upper surface of the mass thus formed is worked into a concave form *b, b*, exhibited in section in Fig. 3, the concavity being bordered by a ledge of two or three inches in width all round save at the base of the oval, where it is much wider. In this wide part of the ledge or *breast* are bored holes, *c, c, c,* for the purpose of drawing off the fused litharge by means of a temporary channel, termed a "gate," d, scooped in the substance of the cupel as occasion requires.

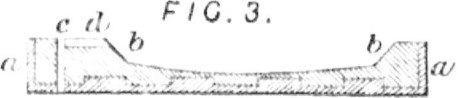

According to the amount of silver lead alloy to be worked at a time the metal ring is from two to four feet long and eighteen to thirty-six inches wide, by four to six inches deep, the thickness of bone-ash in the thinnest part being an inch or a little more. The cupel is set in brickwork under the arch of a small reverberatory furnace, represented in plan in Fig. 4, and in section

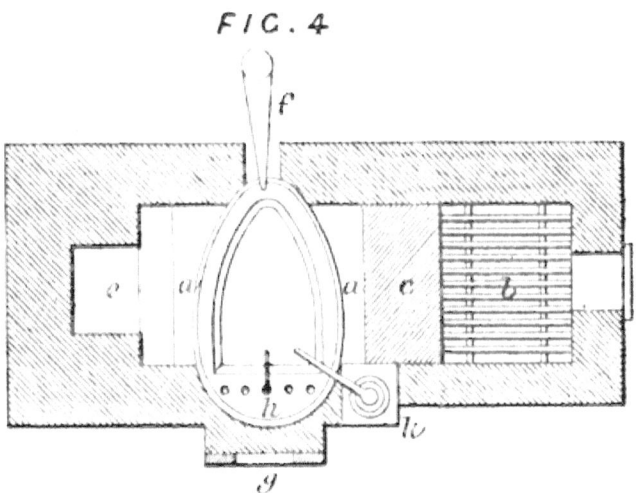

in Fig. 5, a, a, cupel: b, fireplace; c, bridge over which the flame from the fire passes, being reflected downwards, as it were, by the arch d, and passing to the chimney by the flue e, e, running finally underground. The oxidation of the lead is greatly quickened by blowing a stream of air over the fused metal from the nozzle *f*, whilst the operation is watched from time to time through the door-hole opposite to the nozzle *g*.

FIG. 5.

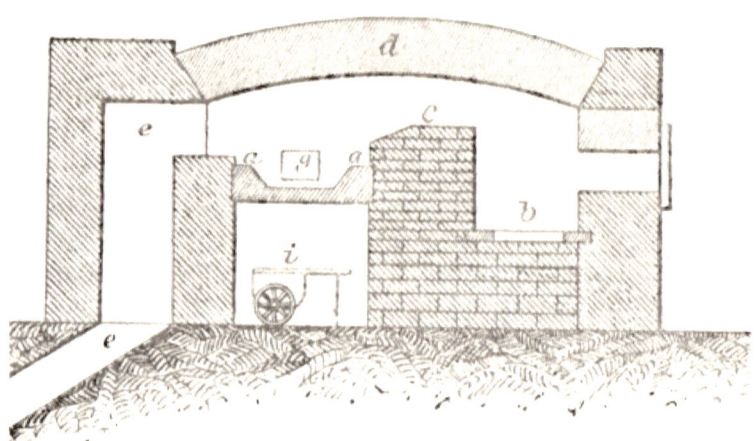

As the lead oxidizes and the fused litharge is removed by the gate *h* (running through the perforations in the cupel and being received underneath in an iron truck or "bogie" *i*), the level of the metal in the cupel is kept approximately constant by running into the cupel through a shoot or gutter fused argentiferous lead from the subsidiary pot *k*, in which is a supply of metal kept molten by a small fire. In this way many cwts. of argentiferous lead are gradually worked up in the same cupel, and ultimately a mass of molten silver of several hundred ounces in weight is obtained. This is drawn off when the operation is finished through a hole bored in the bottom of the cupel; this hole is then stopped up with a plug of ash and the operation is conducted over again. Usually the finishing of the refining is conducted in a separate cupel, the concentration of the silver by removal of lead, &c., being only carried on up to a certain point in the first cupelling furnace, a somewhat higher temperature being required towards the close of the operation; moreover, the litharge separated at the close of the process is apt to contain a little silver, wherefore it is collected apart and reduced to the metallic state and worked over again as argentiferous lead.

20. When silver is associated with considerable quantities of copper, the separation of the precious metal is often effected by a combination of processes consisting of three stages; first, a ternary copper silver lead alloy is prepared by adding to the argentiferous copper ore galena (if not naturally admixed therewith), and smelting the whole, or by adding argentiferous lead to a poor silver copper alloy already smelted or otherwise extracted. This ternary alloy is cast into thin cakes weighing three or four cwt. each. Secondly, these cakes are submitted to the process of liquation or "sweating," whereby a rich argentiferous lead runs away in the fluid state whilst the copper remains as a spongy skeleton, which retains a little silver; to extract this, the spongy copper is again fused with lead and the alloy sweated a second time. Thirdly, the argentiferous lead containing a little copper, is either used again for the alloy employed in another first sweating so as to concentrate silver therein, and is then treated by the cupellation process, or is directly cupelled without further concentration, if rich enough.

In the case of ores in which copper is absent, or only present in small quantities, but which are very rich in lead (*e. g.* ordinary galena) some ingenious devices are in use for concentrating the small portions of silver present in the lead smelted from such ores, so as to extract it without being obliged to re-oxidize all the metallic lead by cupellation. The best-known process of this kind was invented by H. L. Pattinson, and is hence termed "Pattinsonage." It is based on the fact that if an alloy of lead and silver in which the former metal preponderates be allowed to solidify slowly, crystals form consisting of lead retaining much less silver than the original alloy, whilst the fluid portion is proportionately richer in silver; or in other words, that an alloy of silver and lead melts at a lower temperature than another similar alloy containing less silver, whilst such alloys do not remain permanently commingled but have a tendency to separate one from the other, just as oil and water do after shaking up into an emulsion. In practice the operation is carried out by melting the argentiferous lead in one of the central members of a series of from eight to twelve hemispherical iron melting-pots each mounted in a separate furnace, but all arranged side by side. When melted, the fire is withdrawn or dampered down and the lead allowed to cool, the formation of crystals being facilitated by sprinkling water on the top and well stirring in the crusts thus produced. By means of a perforated ladle the crystals are fished out and drained, the ladle being supported by chains from a crane overhead, so that its contents can be readily transferred to the next pot on the right-hand side. When about-two thirds of the lead has thus been fished out, the yet fluid portion is transferred to the next pot on the left, and there worked up in a similar fashion along with the " bottoms," or fluid residues of previous operations, or what comes to the same thing, with pigs of argentiferous lead of the same degree of richness. The poorer first crystals in every pot in the series are thus continually passed into the next adjacent pot on the right to be worked over again, whilst the richer bottoms continually travel to the left. Finally, in the two end pots there are formed respectively crystals of lead so poor in silver that it will not pay to work them any further, and bottoms so rich in silver that they are more conveniently cupelled than Pattinsonized. These end limits are usually reached when the lead retains no more than about 300 grains of silver to the ton ($\frac{5}{8}$ oz.) and when it contains about two per cent. of silver (some 600 oz. per ton).

Parkes' process for desilverizing lead consists in fusing the argentiferous lead with zinc. From the mixture of metals thus produced the zinc separates on standing, floating up to the top whilst the lead remains at the bottom; or rather an alloy of lead and zinc containing about $1\frac{1}{4}$ per cent. of lead floats up to the top, whilst an alloy of the same metals but containing only about $1\frac{1}{2}$ per cent. of zinc subsides; the lighter zinc alloy is found to contain almost the whole of the silver present. Zinc being a volatile metal, is recovered by distillation from the ternary alloy thus gained, whilst the residual silver lead alloy is cupelled, more lead being added if necessary for the operation. One objection to this process is that the desilverized lead retains zinc which must be removed by a refining process before the lead is marketable; this is accomplished in various ways, one of the most successful of which is blowing superheated steam through the fused metal, whereby the zinc is oxidized and removed (more or less lead being also oxidized).

21. **Wet Processes.** — Among the numerous methods of this kind that have been proposed and worked may be mentioned the process of Augustin for the extraction of silver from copper pyrites; this consists in roasting the ore so as to form oxide of iron and oxide or sulphate of copper with sulphate of silver; the roasted ore is then mixed with common salt and again roasted

so that silver chloride [1] is formed (*vide* § 43); the resulting mass is then powdered and treated with warm brine so as to dissolve out the silver chloride, and from the solution thus obtained the silver is precipitated by metallic copper. A modification of this process was proposed by Percy, consisting mainly in the use of sodium thiosulphate as a solvent for the silver chloride in lieu of common salt, and precipitation of the silver as sulphide by means of sodium sulphide. With some kinds of ores, carefully roasting suffices to convert silver sulphide present into sulphate which is dissolved out (along with more or less of the sulphates of other metals) by means of water, and then treated with copper, &c. so as to throw down metallic silver. In Henderson's process for the extraction of copper from cupriferous pyrites containing but little copper (really a modification of Augustin's process, *vide* § 43), a liquid is obtained containing the chlorides of sodium, iron, and copper, with minute quantities of silver chloride, and much sodium sulphate, &c. To isolate the silver from this solution Claudet employs a soluble iodide whereby silver iodide is precipitated, the copper being then extracted from the mother liquors by precipitation with iron. The silver iodide is then treated with zinc, whereby metallic silver is formed together with zinc iodide, the aqueous solution of which is used to precipitate the silver from a fresh quantity of liquor; the metallic silver thus obtained contains an appreciable quantity of gold. Another process for treating the same liquid has

[1] By virtue of the reactions—

$$Ag_2S + 2O_2 = Ag_2SO_4.$$
$$Ag_2SO_4 + 2NaCl = 2AgCl + Na_2SO_4.$$

If unoxidized silver sulphide be present during the roasting with salt, it also becomes converted into chloride, thus:—

$$Ag_2S + 2NaCl + 2O_2 = 2AgCl + Na_2SO_4.$$

been recently described by Mr. J. Gibb,[1] as worked successfully on the large scale; this consists in treating the liquors with sulphuretted hydrogen (prepared from alkali waste by hydrochloric acid) so as to precipitate about one-sixteenth of the copper present as sulphide. This precipitate is found to contain practically the whole of the silver; it is separated from the rest of the liquor and dried without draining or washing so that a certain amount of soluble chlorides still adhere to it. By calcining this precipitate at a low temperature the copper sulphide is oxidized to copper sulphate, which is removed by washing with water, and silver chloride is left in an impure state. From the copper sulphate solution thus obtained the copper is precipitated by means of iron, as is that from the liquor separated from the sulphuretted hydrogen precipitate.

22. **Amalgamation Processes.**—The amalgamation processes in use for the extraction of silver from its ores differ from the method employed in the case of gold as described above (§ 16), in that the silver usually occurs mineralized by combination with non-metals, and hence has to be set free from its compounds before it can be dissolved by the mercury, whilst the gold being already in the metallic state usually requires no such treatment. With certain classes of ores the decomposition of the silver compound is effected by the mercury itself, so that a much greater loss of this latter metal is brought about than that due to imperfect condensation during distillation, flouring, and the like. In others the natural silver compound is previously transformed by roasting or other treatment into some other compound more readily attacked by the mercury; whilst in others the separation of metallic silver

[1] *Chemical News*, xxxi. p. 165.

is effected by means of cheaper reagents, either employed before treatment with the mercury or simultaneously therewith. One of the rudest and most wasteful of these processes, so far as mercury is concerned, is that still largely employed in Mexico, and known as the " Patio " process, being partly carried on in a *patio* or paved space. The ore containing the silver as sulphide, with more or less chloride, is first stamped and then ground with water to a fine powder in a rude but effective mill known as an *arrastre*. The thin mud thus formed is run into a reservoir where most of the water evaporates. The thick paste thus produced is then laid on the patio forming a circular mass of about a foot in depth and forty to fifty feet diameter, weighing some sixty or seventy tons. Salt is then added to the extent of one-fortieth to one-twentieth of the weight and the whole mass well intermixed by treading by mules. After some twenty-four hours from one to two per cent. of " magistral " is added, and the whole again well trodden in and intermixed. This magistral is an impure mixture of iron and copper sulphates, prepared by partially " weathering " copper pyrites by exposure to the air in a wet state, and then completing the oxidation by gentle calcination with a little salt. Mercury is now squirted on to the mixture in a shower by squeezing it through bags or running in from a sheet, and the whole well intermixed by treading; about six parts of mercury are used for every one part of silver present. A chemical action is thus set up and facilitated by turning over and stirring the heap daily for about four weeks. The ultimate effect of this is to form sulphate of soda from the partial oxidation of the sulphur in the silver sulphide and its reaction on the salt, together with that produced by the mutual decomposition of the iron and copper sulphates with the salt, forming in addition iron and copper chlorides. Much mercury is lost by conversion into chloride and also by "flouring;" the residual mercury dissolves the liberated silver, and is finally separated from the mud and metallic salts by a stream of water. The progress of the oxidation is tested daily, and more magistral added to increase the action if necessary, or lime added to decompose the metallic salts and retard the action if the mass heats too much, as in that case the loss of mercury becomes greater. Finally, the amalgam is placed in large canvas bags holding nearly a ton; much fluid amalgam is thus forced out by weight alone; the residual mass is then squeezed and moulded into wedge-shaped solid masses, which are then piled on an iron plate with a hole in it so as to form a circular kind of dome; over this an iron bell ("capellina") is dropped and its edges luted to the iron plate; burning charcoal is then placed in a temporary cylindrical brick furnace built round the bell, so that the mercury is distilled off from the amalgam, the vapour passing through the hole in the bottom plate into a tube dipping into water in a suitable receiver.

23. In what is termed the "Saxon," or Freiberg process, the silver ores are mixed with pyrites containing a little copper (if not already containing such an admixture naturally), and with common salt, and then gently roasted so as to form silver chloride and sodium sulphate, with iron and copper chlorides (§§ 21 and 43). The silver chloride thus formed is reduced to the metallic state by placing the roasted mass in a stout barrel with water and lumps of metallic iron, the barrel being rotated for some hours; mercury is then put in and the whole again rotated slowly for nearly a day. Ultimately an amalgam of silver is produced, from which the earthy matters are washed away, and the fluid mercury separated by squeezing; the

solid mercurial alloy is then distilled.[1] Various modifications of both the Saxon and patio processes are introduced in certain localities according to the special character of the ores used. At Halsbrücke, where the Saxon process is chiefly employed, silver ores are used containing several metals, notably antimony, bismuth, arsenic, copper, iron, lead, zinc, and sometimes nickel, and cobalt; consequently the crude silver left on distilling the amalgam is very impure. On the other hand the "plata pina," from the Mexican capellina distilling process, is frequently almost fine silver, the impurities averaging about six parts in the 1000 (Makins); whilst the silver from other sources is of intermediate character according to the ore employed.

24. In order to obtain pure, or, as it is usually termed, "fine" silver, from the impure crude product, refining by cupellation is usually adopted. The metal to be refined is melted along with lead and the whole cupelled; as the lead oxidizes the other metals are also oxidized and removed, so that finally almost chemically pure silver is obtained, If, however, gold or platinum were present originally these metals are contained in the refined silver. Should gold be present, the high value of this metal makes it worth while to extract it by wet processes, consisting of boiling the auriferous silver with nitric or sulphuric acid which dissolves the silver; if only one part of

---

[1] By virtue of the following actions: the metallic iron converts the copper chloride present into spongy metallic copper—

$$Fe + CuCl_2 = FeCl_2 + Cu;$$

this reacts on the silver chloride setting free silver and reproducing a copper chloride (cuprous chloride for the most part).

$$2AgCl + 2Cu = 2Ag + Cu_2Cl_2.$$

If the mercury be added before the silver has been reduced to the metallic state by the first rotating of the barrel and contents, much mercury is lost owing to its conversion into chloride by taking the place of the metallic copper in this second reaction.

gold in 2000 be present its separation can be profitably conducted. Of the two acids, sulphuric is to be preferred, not only on the ground of its greater cheapness, but also because the silver is more completely dissolved out and a fine gold is thus obtained; indeed, gold refined by nitric acid will often part with a little silver to sulphuric acid. The separation is in practice carried out by fusing together various kinds of auriferous silver and gold containing small quantities of silver, so as to obtain an alloy containing about one quarter of its weight of gold (whence the term *quartation*, often applied to this separation process); this is then granulated and boiled (preferably in a platinum vessel) with the solvent acid until all the silver is dissolved out. From the silver salt thus formed the silver is regained either directly by precipitation with plates of copper, or by addition of a soluble chloride and reduction of the silver chloride thus thrown down. The latter method yields the finer silver, especially if the auriferous alloy employed contained baser metals in small quantities; in this case, however, Miller's chlorine process (§ 17) is more satisfactory as a means of separating fine gold.

25. **Platinum.**—This metal has been known for nearly a century and a half, being originally found in South America and named after silver (*plata*) of which it was supposed to be an impure ore. As experience soon taught that this ore was quite unworkable by ordinary metallurgical processes, it was disregarded until the researches of Wood, Berzelius, and Vauquelain, and more especially of Wollaston, and subsequently of Deville and Debray, demonstrated its peculiar properties and their value for numerous purposes, and showed how the metal could be practically worked. Besides occurring in South America platinum is found native but associated with small quan-

tities of somewhat analogous but less used metals, in the Oural mountains and valleys, and in California, Mexico, San Domingo, and Australia. It occurs either as nuggets sometimes weighing upwards of 20 lbs. or as small grains somewhat after the fashion of gold, and is generally found in the alluvial deposits. The accompanying metals, palladium, iridium, osmium, rhodium, and ruthenium, are more or less completely separated from the platinum by processes essentially of one or other of the following three kinds : either the ore is fused by the aid of an oxyhydrogen flame on a bed of lime, by which means palladium and osmium are volatilized, and a fused alloy of the other metals left in which platinum largely predominates, iridium being present only in small quantities, and rhodium and ruthenium only to a yet lesser extent; or the platinum is separated by a wet process due to Wollaston ; or an alloy of lead, platinum, and small quantities of palladium, is formed by fusing the platinum ore with galena, litharge, and a flux of pounded glass, the other metals being undissolved by the newly reduced lead (Deville and Debray); from this alloy the lead is removed by oxidation in hot air, after the manner of cupellation, an oxyhydrogen flame being used, the palladium present being thus also expelled.

Of these three processes, the first yields an alloy which is well adapted to the manufacture of crucibles and other vessels intended to resist high temperatures and the action of chemicals, as it is less fusible than platinum and more resistant as to corrosion ; but it is not so well suited for the manufacture of large vessels such as sulphuric acid concentrators, in which the junctions are fused together by a skilful application of an oxyhydrogen blowpipe (autogenous soldering). Deville and Debray's process also is apt to yield a less pure metal than Wollaston's, owing to the difficulty in completely removing by subsidence the particles of the other metals undissolved by the molten lead. Hence for certain special purposes (such as the resistance coils of Siemen's pyrometer, § 84), platinum thus prepared cannot be employed even though it have been freed from palladium and osmium by means of the oxyhydrogen blowpipe.

26. Wollaston's process, by which absolutely pure platinum can, with due care, be obtained, essentially consists in treating the platinum ore with nitric acid, after washing with hydrochloric acid for the purpose of removing any baser metals, magnetic iron ore, &c., present; the latter can to a large extent be separated by a magnet. A mixture of nitric and hydrochloric acids, *aqua regia*, is then allowed to act on the purified ore for some hours or days, certain proportions and strengths being usually chosen, so as to prevent as much as possible the solution of iridium, which mostly remains undissolved in combination with osmium and a little rhodium and ruthenium, forming white metallic scales termed *osmiridium:* about three parts of hydrochloric acid of sp. gr. 1·20, and one of nitric acid of sp. gr. 1·35, will answer well, or analogous mixtures containing about the same percentages of anhydrous hydrochloric and nitric acids. The acid solution of platinum is then treated with sal-ammoniac, whereby platino-chloride of ammonium is precipitated, iridium, rhodium, and palladium being retained in solution. To avoid the possible contamination of the precipitate with iridiochloride of ammonium, it is well washed with water and pressed; the latter salt being much more soluble than platino-chloride of ammonium is thus removed. Finally the precipitate is strongly heated, whereby a spongy mass of platinum is left;[1] this is made into a paste

[1] The following equations represent the production of ammonium platino-chloride and its decomposition by heat. By the action of

with water and pressed into moulds, which are slowly dried and gradually heated to whiteness and then forged into compact masses ; or the spongy platinum is pressed into cakes and fused in a lime crucible by the oxyhydrogen flame. From the mother liquors of the platinochloride of ammonium more of that salt can be obtained by precipitating all metals by a plate of zinc, redissolving in aqua regia and adding sal-ammoniac ; this precipitate is apt to contain iridium. From the final mother liquor palladium is separated by neutralizing with sodium carbonate and adding mercury cyanide solution, whereby palladium cyanide is thrown down from which metallic palladium in a spongy state is obtained by simply heating. The mother liquors of this precipitate contain rhodium : from the osmiridium left undissolved by the aqua regia in the first instance, the metals osmium and iridium with smaller quantities of rhodium and ruthenium are separated by special processes.

Fig. 6 represents one simple form of Deville and Debray's oxyhydrogen furnace for the fusion of platinum, &c.; the receptacle for the metal to be fused is a compact mass of lime or marble (which immediately becomes converted into quicklime on the inner surface by the heat) hollowed out into a dish shape ; this is covered

the chlorine set free from the aqua regia, platinum chloride is formed—

$$Pt + 2Cl_2 = PtCl_4.$$

On adding sal-ammoniac to this, combination occurs and ammonium platino-chloride results—

$$PtCl_4 + 2NH_4Cl = Pt(NH_4)_2Cl_6.$$

Finally, this compound breaks up on heating, forming sal-ammoniac, chlorine, and platinum—

$$Pt(NH_4)_2Cl_6 = Pt + 2Cl_2 + 2NH_4Cl,$$

the sal-ammoniac being to some extent further decomposed by the chlorine in virtue of a secondary reaction.

by a similar block in the crown of which is a perforation by means of which the oxyhydrogen flame is made to play inside the crucible furnace ; for large masses several distinct flames may be employed. The most convenient form of oxyhydrogen burner for this purpose is simply an ordinary blowing lamp, *i.e.* two tubes arranged concentrically, the combustible gas passing into the annular space

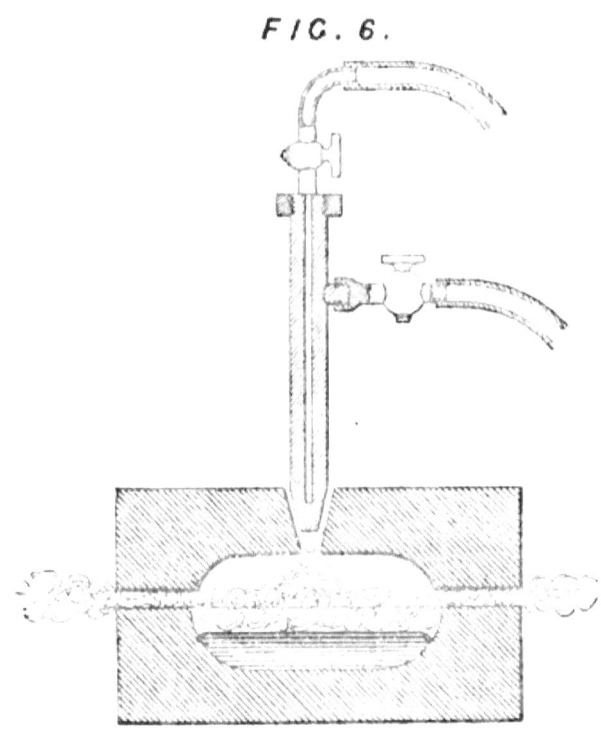

FIG. 6.

outside, and the blast of air or oxygen being passed through the interior tube, the supply of each gas being regulated by cocks. A large flame of this kind supplied with oxygen and hydrogen (best from cylinders of the compressed gases) enables many effective combustion experiments to be made : a steel carving-knife burns with great brilliancy in the flame, and if a large plate of thin sheet iron be held in a vertical position, and the flame cautiously moved over its flat surface, writing may be roughly traced by the perforation of the plate thus brought about (Fig. 7). With even a small flame, such as that used for the lime light, thin platinum foil may be similarly perforated; for this purpose it is not necessary to use

FIG. 7.

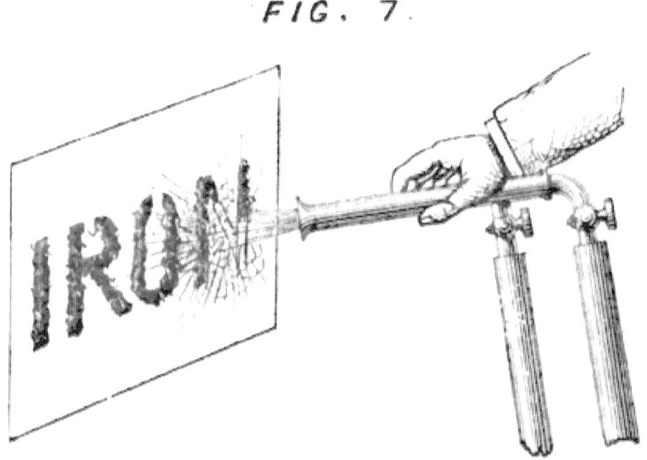

hydrogen gas, as ordinary coal gas will often produce the same result; carbon oxide and oxygen, also, will suffice, the heat evolved in the oxidation of carbon oxide being superior to that produced during the combustion of hydrogen, a given quantity of oxygen being employed. The oxy-carbon oxide flame, however, is less compact than that of oxyhydrogen owing to the combustion taking place with less rapidity; in the same way a large jet of carbon oxide ceases to burn in the air when the velocity with which the gas issues exceeds a certain amount, the flame as it were blowing itself out, the rapidity of propagation of combustion through the mixture of carbon oxide and air formed at the orifice of the jet being very small as compared with the similar velocity in the case of hydrogen.

27. **Mercury.**—Although of considerably less intrinsic value than the preceding metals, mercury is classed along with them on account of its not rusting in the air at ordinary temperatures: it occurs native both as silver amalgam, and as almost chemically pure metal exuding in drops from the crevices of the lodes of *cinnabar*, the most important ore of mercury, consisting of its sulphide, from which the main supplies of the metal are obtained. The cinnabar mines of Almaden in Spain have been worked for considerably upwards of 2,000 years; for Pliny states that the Greeks imported from thence red cinnabar to be used as a pigment 700 years B.C., and that in his time large quantities were brought to Rome from the same source. At present these mines produce about 1,000 tons of metallic mercury yearly. At Idria in Transsylvania large quantities of mercurial ore are raised, about 150 tons of metal being annually extracted; owing to the imperfect apparatus employed much waste is occasioned in the extraction, so that much larger amounts of metal could be produced from the same quantity of ore; it is stated that as much as 600 tons could be extracted yearly, but that the amount of produce is purposely limited to about 150 tons in order to keep up the price. Cinnabar has also been found in large quantities in California, China, Japan, and Peru.

The most antique process for the isolation of the metal, still largely employed at Almaden, consists in simply heating the ore in contact with air so as to oxidize the sulphur and volatilize the mercury,[1] the latter being condensed in rudely constructed series of globular tubes,

---

[1] By virtue of the reaction $HgS + O_2 = Hg + SO_2$.

termed *Aludels*, Fig. 8, inserted one within the other. The oxidation of the ore is effected in what is styled a *Butyrone* furnace; the ore is placed in the upper portion of a doubly arched vault, the lower arch being perforated with holes; into the lower portion wood is introduced through a door and kindled, most of the smoke escaping by a side chimney. The heat from this fire by and by causes the sulphur of the cinnabar to burn, and this combustion furnishes sufficient heat to cause the rest of the operation to go on spontaneously. The sulphur dioxide formed, and the mercury vapours pass out at the top into a series of aludels laid on a double slope: the centre aludel is perforated so that the mercury condensed in the aludels can flow out into

FIG. 8.

a gutter. Finally, the uncondensed vapours are led into a wide vault with a trough containing water at the bottom; in this more mercury condenses, whilst the sulphur dioxide escapes at the far end. Twelve series of aludels are applied to each furnace; and as there are about fifty aludels in each series, the number of joints and consequent loss by leakage is very great. The joints are luted with loam; after each operation the whole series of tubes is taken to pieces, and the condensed mercury poured out of each aludel separately. An enormous amount of labour is thus rendered requisite, whilst the leakage of mercurial vapours into the air exerts a most prejudicial effect on the health of the workmen. At Idria, the aludel condensers have been replaced by series of masonry chambers, the rest of the process being much the same; even here

CHIEF INDUSTRIAL APPLICATIONS.

large leakage of mercurial vapours and loss of metal is apt to occur.

28. A much more satisfactory method of proceeding is adopted in other works; the ore is mixed with lime and heated in proper distilling vessels, when mercury, with a little sulphide and oxide, distils over.[1] An old form of apparatus for this purpose, known as the "Palatinate gallery," is represented in Fig. 9: two double rows of earthen or iron cucurbits, a, a, a, a, are mounted in a flue, or elongated fireplace, of which b represents the

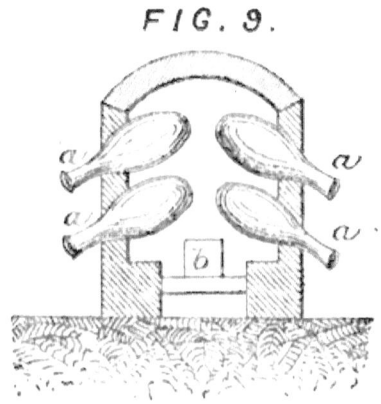

FIG. 9.

bars. Each cucurbit is capable of holding 60 to 80 lbs. of a mixture of ore and from a third to a quarter of its weight of lime; from 32 to 52 cucurbits are mounted in one gallery. The heat expels mercury as vapour, which

---

[1] The chemical change in this operation is somewhat more complex than in the butyrone furnace. The main action is represented by the equation—

$$4HgS + 4CaO = 3CaS + CaSO_4 + 4Hg,$$

calcium sulphide and sulphate being formed. The little amount of mercurous oxide that sublimes may be regarded as formed thus—

$$8HgS + 8CaO = 7CaS + CaSO_4 + 4Hg_2O.$$

is condensed in stoneware receivers applied to the cucurbits and kept half full of water; after the operation is finished these are emptied out into a tub; the black powder (sulphide and oxide of mercury) washed out from the receiver by the water is collected and heated with lime over again. At Landsberg in Bavaria a superior distillation apparatus is used constructed after the fashion of a bench of gas retorts by the late Dr. Ure;[1] whilst in other places arrangements are employed very similar to the capellina distillation apparatus used for the separation of mercury from gold and silver amalgams (§§ 16 and 22).

Mercury as met with commercially often contains lead, tin, and other cheaper metals, fraudulently added to increase its weight; these cannot always be wholly removed by redistillation, and hence must frequently be separated by chemical means, such as digesting the impure mercury with dilute nitric acid, mercurous nitrate solution, or strong sulphuric acid, when the foreign metals dissolve, their solution being probably promoted by the galvanic action set up.

Of the other metals belonging to the class of noble or precious metals (palladium, iridium, rhodium, osmium, ruthenium), the first and second have been employed for certain special purposes but only to a limited extent; thus palladium, irrespective of its use for certain scientific investigations on account of its peculiar power of condensing hydrogen, forms a valuable alloy with silver, useful for dental purposes, mathematical instruments, &c.: whilst an alloy of platinum and iridium also possesses valuable properties, being less fusible and less attacked by corrosive agents than even platinum itself (*vide* § 25).

---

[1] *Vide* Ure's *Dictionary of Arts, Manufactures, and Mines*, Art. "Mercury."

CHAPTER III.

METALLURGY OF THE MORE IMPORTANT BASE (READILY OXIDIZABLE) METALS.

29. **Iron.**—Without question iron is the metal *par excellence* which could least be spared by civilised nations; invaluable as a means of currency and for other purposes as gold is, its uses are far outweighed by those of iron, a fact recognized ages ago in more senses than one when the Eastern sage predicted his downfall to Crœsus, saying after having viewed his treasures, " Know, O King, that he who possesses more iron will soon become master of all this gold." Most of the practical importance of iron is due to the very different properties possessed by the pure, or nearly pure, metal (*malleable iron*), and by the substances formed by its union with small quantities of carbon (*cast iron* and *steel*);||the former from its fibrous character and consequent great toughness, strength, and tenacity, and its power of becoming plastic and welding together, and hence being readily shaped, at a high temperature, is applicable to purposes for which no other substance could be employed; whilst the comparative fusibility of cast iron, conjoined with its considerable strength, and the hardness and elasticity of steel when subjected to certain alternations of temperature, render these

substances indispensable to the mechanical and civil engineer, as well as to almost every handicraftsman.

The following table gives a general idea of the composition of the more important iron ores :—

	Hæmatite.	Brown Ore.	Magnetic Ore.	Spathic Ore.	Clay Ironstone.
Ferric oxide, Fe_2O_3	90—100	40—70	30—70	0—3	0—3
Ferrous oxide, FeO	—	0—5	13—33	36—50	35—55
Alumina, Al_2O_3	0—2	1—7	0—5	0—2	1—7
Lime, CaO	0—3	1—7	0—5	0—4	1—14
Magnesia, MgO	0—1	0—2	0—2	0—4	1—9
Oxide of manganese, MnO	0—1	0—3	0—1	1—25	0—2
Silica, SiO_2	0—10	1—35	0—10	0—5	2—17
Carbon dioxide, CO_2	0—1	0—5	0—10	37—42	22—37
Phosphoric anhydride, P_2O_5	0—1	0—2	0—2	trace	0—2
Sulphuric anhydride, SO_3	0—1	trace	trace	trace	0—3
Water, H_2O	0—1	6—18	0—4	—	0—1
Essential Composition.	Ferric oxide with little or no earthy & siliceous matters.	Hydrated ferric oxide with more or less earthy and clayey matters.	Ferric and ferrous oxides with small quantities of earthy matters.	Ferrous carbonate and manganese carbonate; crystalline.	Ferrous carbonate and generally large quantities of earthy and clayey matters.

Native iron, presumably of meteoric origin, is so comparatively scarce as not to be of importance as a source of the metal; still tools have been met with among semi-civilised nations fashioned from iron derived from this source. The leading ores of the metal are the *oxides*, especially the *magnetic oxide* (which derives its name from the fact that the natural loadstone is a variety of this substance); the *peroxide*, occurring as the minerals hæmatite, specular iron ore, &c.; and the *hydrated peroxide*, forming brown iron ore, iron ochre, &c.; the *carbonate*, forming spathic iron ore when crystallized, and clay ironstone when more or less intermixed with earthy matter ; and the *sulphides*, especially *pyrites*. The last, however, is of more importance, industrially, as a source of sulphur than as an iron ore, although by the processes already alluded to (§ 10; *vide* also § 43) for extracting copper from cupriferous pyrites, a peroxide of iron can be thence prepared from which metallic iron can be readily obtained, and which indeed is actually made into a source of that metal in the puddling furnace (§ 38).

30. From these and analogous ores metallic iron in a state of more or less purity is obtained by processes which may for practical purposes be divided into two classes, viz., those where tolerably pure iron (or steel) is obtained at one operation (*direct processes*) ; and those where a very impure iron is first prepared, containing several per cents. of carbon with other substances, notably sulphur, phosphorus, and silicon, which greatly add to the fusibility of the product ; from this " pig iron " or " cast iron " malleable (or nearly pure) iron is subsequently obtained by the aid of refining and purifying processes. These two methods really insensibly shade into one another, as the substances prepared by direct processes often only differ from pig iron in containing smaller quantities of impurities, having, indeed, characters intermediate between those of malleable iron and of pig iron. When these impurities are limited to carbon in quantity not much exceeding about 1 per cent. (sulphur, phosphorus, and silicon being absent, or occurring only in traces) the substance possesses the peculiar properties of *steel*, *i.e.*, it is far harder than wrought iron, especially when heated and suddenly cooled ; it further possesses a peculiar elasticity and power of retaining a sharp edge which renders it especially valuable for cutting instruments and tools generally.

The following table illustrates the general chemical composition of malleable iron, steel, and pig iron :—

	Malleable Iron.	Steel.	Pig Iron.
Iron	99·0—99·5	98.5—99·5	90·0—95·0
Carbon	0·1—0·5	0·5—1·5	2·5—4·0
Silicon	0—0·2	0—trace	0·1—3·5
Sulphur	trace	0—trace	trace—0·5
Phosphorus	0—0·5	0—trace	trace—1·5
Manganese	0—trace	0—trace in cementation steel; 0·1—2·0 and more in Bessemer and analogous steels	trace—2·0
Characteristics	Welds readily, comparatively soft, very difficultly fusible, will not harden, very tough and tenacious.	Can be welded more or less easily, less infusible than malleable iron, can be hardened and tempered.	Will not weld, comparatively readily fusible and easily cast, will not harden like steel, more brittle than tempered steel or malleable iron.

31. By processes of the direct kind, when properly worked, substances are obtained which are of the nature of malleable iron, or soft steel; one of the oldest forms of apparatus used for this purpose is known as the " Catalan Forge " (Fig. 10); in principle this is much of the nature of a blacksmith's forge, a stream of air being forced through a nozzle, or " tuyere," into a charcoal fire in which lumps of iron ore to be reduced are placed, a pile of ore and charcoal being raised over the fire : to get good results a hæmatite or magnetic oxide of considerable purity must be employed; according to the way in which the manipulation is effected, the substance reduced in the furnace will vary in character from nearly pure iron to soft steel: the spongy mass of metal produced is hammered and forged into bars whilst still red-hot. A rough

form of bellows-furnace of this description has been used in India from time immemorial, producing a metal from which steel of the finest quality (Wootz) is prepared by melting or fritting the partially carbonized iron thus produced with wood in closed crucibles. Direct processes,

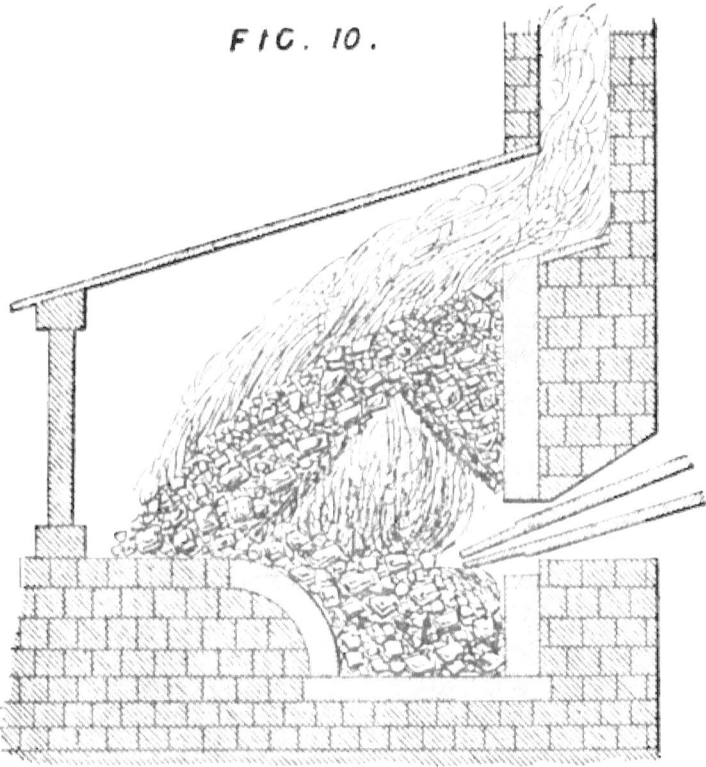

FIG. 10.

however, until of late years, have comparatively fallen into desuetude; during the last decade several attempts have been made to revive them in a more perfect form, so as to obviate several objections to which the usual roundabout methods of blast-furnace smelting, with subsequent refining and puddling processes or Bessemerizing, are open. Whilst it cannot be said that any of these have as yet in any considerable degree superseded the blast furnace, there is every reason to believe that such a result will ultimately be brought about to a greater or less extent ; the gas producers, "regenerative furnaces," and other apparatus constructed by Siemens affording the means of overcoming many practical difficulties that have hitherto stood in the way of direct processes being successfully carried out, so that iron and steel can be readily produced direct from the ore on a large scale.[1] The regenerator (so called from its introducing again into the furnace the heat thence escaping in the exit gases), enabling very high temperatures to be produced with a comparatively small expenditure of fuel, renders it possible to effect many kinds of operations which, with ordinary furnaces, are either impracticable or too costly from the amount of fuel consumed : it consists simply of vaults, or flues, in which are stacked piles of firebricks with interstices between, so that as the spent gases and flames from the furnace pass through, the bricks become heated up, and the waste products of combustion finally issue at the far end to the chimney almost wholly robbed of their heat. Two pairs of regenerators are usually employed, the waste gases passing through one pair and heating it up, whilst the gaseous fuel from a gas producer and the air to burn it are passed inwards to the furnace through the other pair previously heated ; by a suitable arrangement of valves and doors the gases are made to pass alternately through one and the other pair of regenerators, the fuel and blast being coincidently shunted to the pair just heated up ; in this way almost all the heat generated in the furnace is again brought back to it in the heated blast and gas, so that the practical limit to the tem-

[1] *Journal of the Chemical Society*, 1873, p. 661.

III.] CHIEF INDUSTRIAL APPLICATIONS. 59

perature attainable is simply the fusibility at intensely high temperatures of even the most refractory construc-

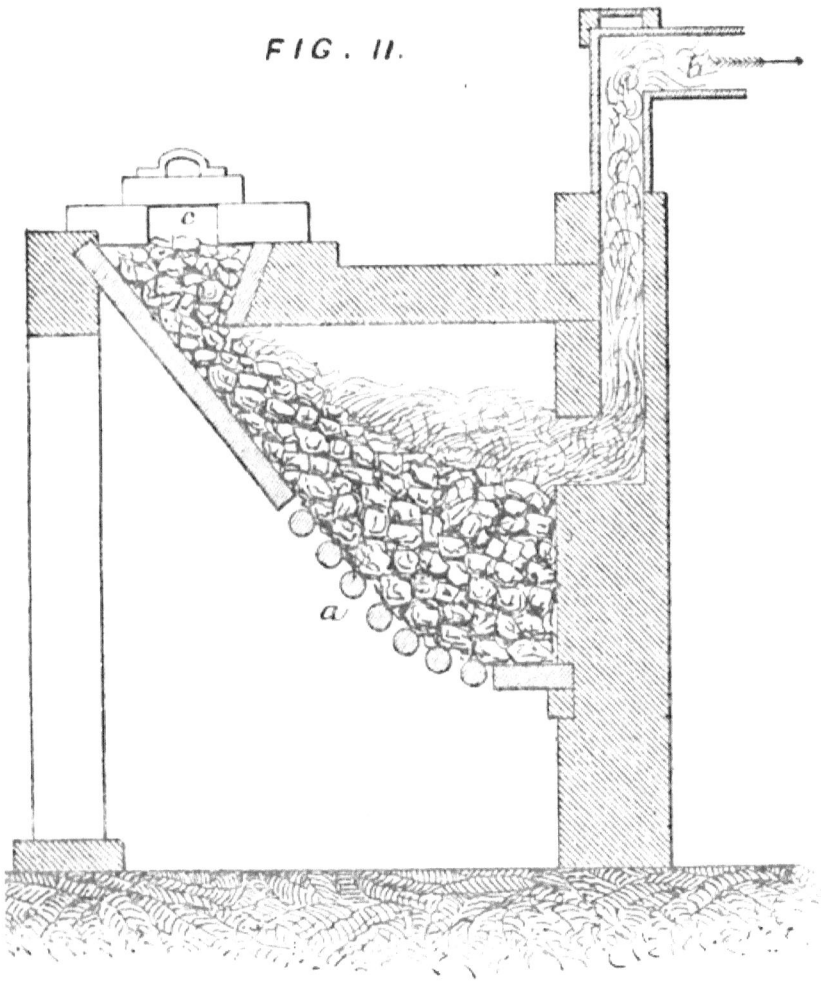

FIG. II.

tive materials obtainable. The combustible gases which serve as fuel in these furnaces are produced by the arrangement shown in Fig. 11; air is admitted through the

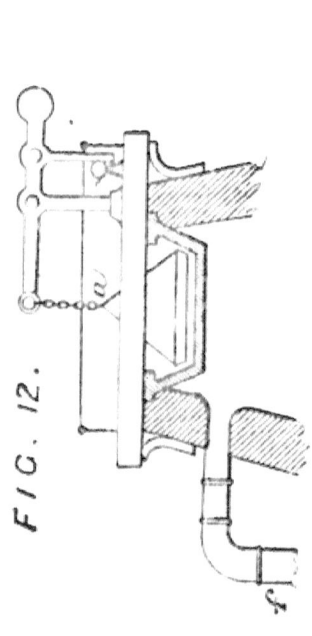

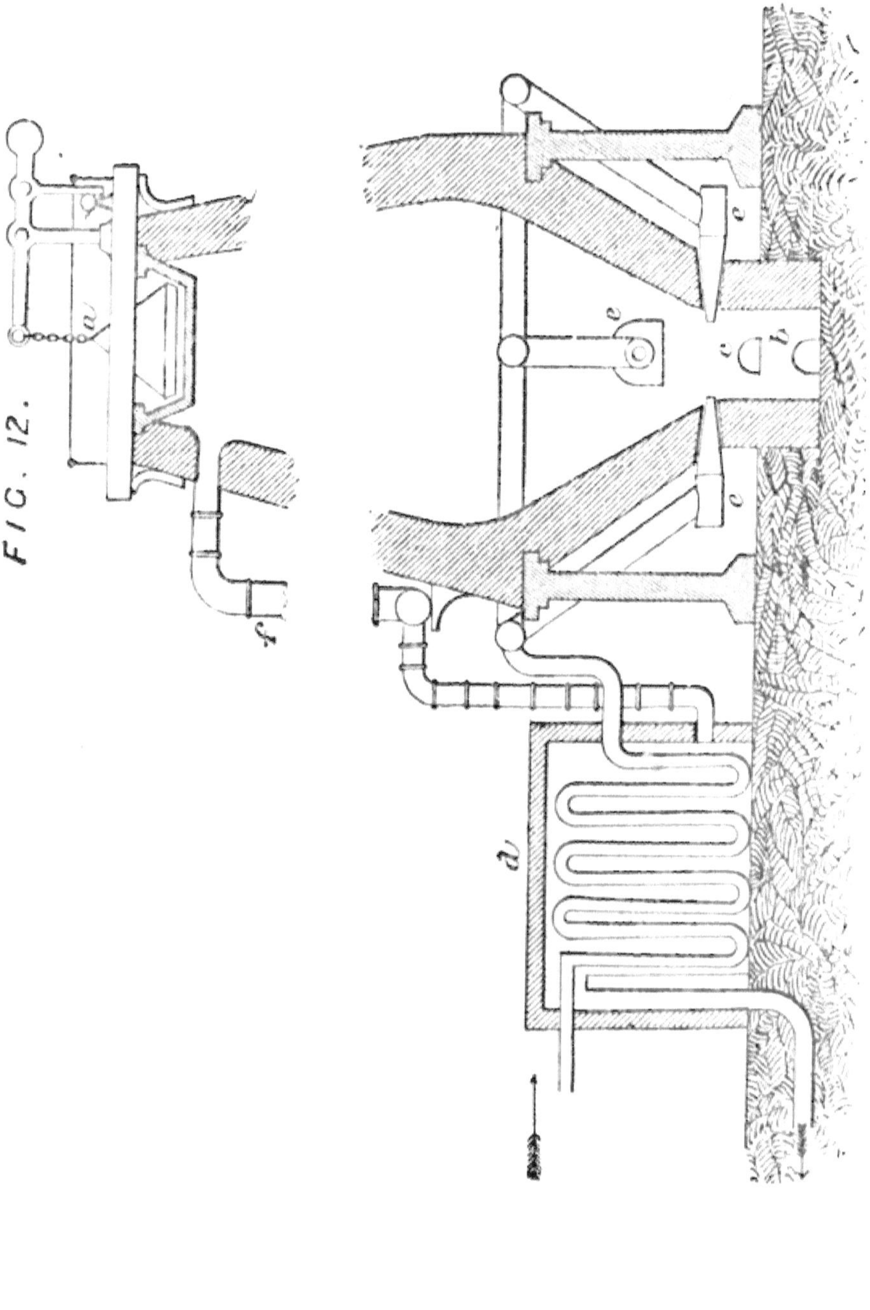

Fig. 12.

grate, a, and as the first products of combustion pass through the heated mass of fuel above the fire bars, all or nearly all the carbon dioxide present becomes converted into carbon oxide, so that the gases which issue at the gas flue, b, consist of nitrogen, carbon oxide, and more or less carburetted hydrogen and hydrogen, according to the nature of the fuel burnt, &c. : thus Siemens gives the following analysis of the gas given by a mixture of 3 parts caking and 1 part non-caking coal :—

Combustible gases.	Carbon oxide	24·2 per cent. by vol.			= 34·6
	Hydrogen	8·2	,,	,,	
	Carburetted hydrogen	2·2	,,	,,	
Incombustible gases	Carbon dioxide	4·2	,,	,,	= 65·4
	Nitrogen	61·2	,,	,,	

100·0

As the fuel sinks in the gas producer, fresh substances are introduced through the hopper, c, ashes being stoked out through the fire bars. One great advantage of this gas producer is that many kinds of refuse, of no value at all as fuel when burnt in the ordinary way, can be readily utilized and transformed into valuable gaseous fuel : whilst the gases thus produced are more convenient as fuel for many manufacturing operations than coal and such-like substances.

32. By far the greater portion of iron employed industrially, however, is extracted from the ores used by the aid of the "blast furnace." Fig. 12 gives a notion of the usual form of this gigantic metallurgic appliance, 70 to 80 feet and upwards being a usual height. Virtually the blast furnace is a vertical tube swelling out in the middle or rather lower down (the *boshes*) ; at the top end of this tube the materials employed are continually inserted, whilst at the bottom the products of the chemical actions

taking place inside are withdrawn continuously or periodically. The ore to be smelted, coke, coal, or charcoal to reduce it, and limestone or other analogous material to serve as a flux for the earthy matters present, enter at the top, whilst pig iron and a fused earthy siliceous mass known as "cinder," pass out in the fluid state at the base through arched holes, b and c, in the thick masonry of the lowest portion of the furnace, termed respectively the "tapping hole" and the "cinder hole." In addition, air, usually heated (either by an arrangement analogous to Siemens' regenerator, or by a superheater, like that represented in the cut, d), is continually forced in by powerful blowing engines at the bottom through tuyeres (*hot-blast*); whilst the gaseous products of combustion and of the reduction of the ore in the furnace escape at the top. Usually the top is covered in by a cone or bell of iron, a, capable of being lowered by a counterpoise and winch when required to introduce materials; when raised, the hot exit gases pass through the side tube, f, by which they are led to fireplaces under boilers, &c., and burnt so as to generate steam, or to heat the superheaters for the hot-blast, &c.; for the issuing gases invariably contain much carbon oxide, wherefore much more fuel is requisite to smelt iron in a blast furnace than would, theoretically, be required if the carbon of the fuel were burnt wholly to carbon dioxide.

33. Usually the ores employed are subjected to a preliminary calcining so as to convert ferrous carbonate into ferric oxide, to expel water, and generally to open out the texture of the ore; calciners much resembling an ordinary limekiln are used for this purpose; if coal be employed as fuel it is coked by the heat in the upper part of the furnace, so that what descends to the tuyeres and is there burnt by the entering hot-blast is always

CHIEF INDUSTRIAL APPLICATIONS. 63

carbon: this carbon is either burnt directly to carbon oxide, or else if burnt firstly to carbon dioxide, is immediately converted into carbon oxide by the white-hot carbon, inasmuch as little or no carbon dioxide exists in the gases inside the furnace at the tuyere level (Lowthian Bell). Virtually, therefore, the air becomes immediately converted into a mixture of about 2 volumes of nitrogen and 1 of carbon oxide which ascends through the furnace, producing a number of complex changes as it comes in contact with the ore, the end results of which may be thus summarized.[1] At the top of the furnace the ferric

---

[1] *Vide* Lowthian Bell, *Chemical Phenomena of Iron Smelting*; also Alder-Wright, *On the Chemical Changes accompanying the Smelting of Iron* (Lecture at the Royal Institution, Friday, March 13th, 1874). The actual chemical changes which take place in the blast-furnace are indicated by the following equations (so far as the carbon oxide and iron oxide are concerned):—

[A] Reduction of higher oxide to lower oxide and to metal by gaseous carbon oxide—

(1) $Fe_xO_y + CO = Fe_xO_{y-1} + CO_2$.
(2) $Fe_xO_y + yCO = xFe + yCO_2$.

[B] Oxidation of metal to lower oxide and higher oxide by carbon dioxide [changes the reciprocals respectively of (1) and (2)]—

(3) $xFe + yCO_2 = Fe_xO_y + yCO$.
(4) $Fe_xO_{y-1} + CO_2 = Fe_xO_y + CO$.

[C] Reduction of carbon oxide to carbon—

(5) $xFe + yCO = Fe_xO_y + yC$.
(6) $Fe_xO_{y-1} + CO = Fe_xO_y + C$.

[D] Reaction of reduced carbon on iron oxide forming carbon dioxide—

(7) $2Fe_xO_y + C = 2Fe_xO_{y-1} + CO_2$.
(8) $2Fe_xO_y + yC = 2xFe + yCO_2$.

[E] Reaction of carbon dioxide on reduced carbon—

(9) $CO_2 + C = 2CO$.

At any given point in the furnace all the chemical tendencies thus indicated are at work simultaneously to extents dependent on the temperature and other circumstances; so that the action at that point is due, as it were, to the single resultant of the various forces

oxide is partially reduced to lower oxides, and these react again upon the carbon oxide, setting free carbon in a state of very fine division in the pores of the partially reduced ore : in this part of the furnace the coal (if that fuel be used) is coked, and limestone is calcined to quicklime. If carbonate of iron (raw clay ironstone, &c.) be used instead of calcined ore it becomes here converted into oxide, the same changes further taking place.

In the middle region of the furnace, reduction goes on only to a limited extent, the iron-oxidizing tendencies nearly balancing the iron-reducing tendencies, whilst the carbon-depositing and carbon-oxidizing tendencies are also nearly balanced ; so that the chief effect produced on the ore in sinking through this part of the furnace is that its temperature is greatly raised.

In the lowest portion of the furnace the reduction of the ore is completed, partly by the finely divided carbon contained in its pores, partly by the action of alkaline cyanides (formed by the mutual reaction of carbon, alkaline carbonates from the ore and fuel, &c., and the nitrogen of the blast) :[1] the reduced iron melts, dissolving as it does so a quantity of the finely divided carbon with which it is surrounded ; at the high temperature of the vicinity of the tuyeres, the reduced iron and the carbon, &c., jointly cause the reduction of more or less silicon phosphorus and sulphur (if present), which are also dissolved by the fluxing iron. The earthy matter of the ore, the ash of the fuel, and the lime, &c., added as flux, also melt together forming a kind of impure glass or lava

---

involved, viz., the *iron reducing* tendencies [A] and [D] ; the *iron oxidizing* [B] and [C] ; the *carbon reducing* [C] ; and the *carbon oxidizing* tendencies [D] and [E].

[1] In the case of potassium cyanide by virtue of the reaction—

$$K_2CO_3 + 4C + N_2 = 3CO + 2KCN.$$

(*cinder*);[1] this and the molten pig-iron sink down to the bottom of the furnace together, the latter, being heaviest, sinking to the bottom, whence it is drawn off from time to time through the "tapping hole," into moulds of sand in which it is cast into "pigs;" the cinder flowing off constantly through the "cinder hole," excepting for a while after tapping off the pig-iron, when the cinder inside the furnace falls below the level of the cinder hole.

34. In blast furnaces that have been long in use the amount of alkaline cyanides present internally is something astonishing, the small quantities of alkalies continually brought into the furnace accumulating therein until relatively enormous amounts are present: this accumulation is apparently brought about as follows; the alkalies are volatilized to a considerable extent (query, in the metallic form as sodium, potassium, &c. ?) in the intensely hot lower portion of the furnace, and in this condition are largely converted into cyanides by the conjoint action of the finely divided deposited carbon and the nitrogen gas present: these cyanides are disseminated through the gases, and on passing upwards to cooler regions are deposited (along with dust, &c., mechanically carried up) as a whitish sublimate, or "fume," on the surface of the ore, &c.; the ore in descending thus becomes largely impregnated with cyanides, and when the mixture reaches the lower part of the furnace these cyanides take up oxygen, completing the reduction of the ore and forming cyanates which are immediately decomposed by the heat; the alkali metals of the decomposed

---

[1] The term "slag" is often applied to this material; but "slag" more appropriately refers to the somewhat analogous substance beaten out of the blooms during shingling after puddling (§ 38), the name being derived from the German *schlagen*, "to beat"; "cinder" (query, from *sintern*, "to trickle"?) properly refers to the lava-like mass that runs from the blast-furnace and other analogous furnaces.

cyanates become again transformed into cyanides which are again filtered out of the gases in the upper parts of the furnace, and so on. To so great an extent does this accumulation of alkalies take place that as much as four cwt. of alkali-metals and two cwt. of cyanogen per ton of iron run have been found in the gases near the level of the tuyeres (Lowthian Bell). This formation and accumulation of alkaline cyanides serves to explain two anomalies in the working of the blast-furnace : first, on starting a newly-built blast-furnace, much more fuel is required during the first few days than after some weeks' working, even after allowance is made for the fact that a considerable amount of extra fuel is required to furnish the heat communicated to the masonry of the furnace ; direct observation shows that the cyanides present increase, by virtue of this accumulative process, *pari passu* with the diminution in amount of extra fuel required during this period. Secondly, when analyses are made of the gases [1] present at different levels of the furnace, and the ratios calculated between the weights of nitrogen and carbon and nitrogen and oxygen, contained therein, it is observed that whilst at every level there is more oxygen and more carbon per 100 of nitrogen than that due to the blast alone (assuming the oxygen of the air to be wholly converted into carbon oxide, which gives a maximum of carbon as present in the gases) the carbon and oxygen

---

[1] It deserves notice that in the well-known analyses of blast-furnace gases, made some years ago by Bunsen and Playfair, *cyanogen gas* is represented as being present therein; the author's experience on the subject, however, leads him to the conclusion that the substance viewed as cyanogen was really *hydrocyanic acid vapour* caused by the decomposition—by carbon dioxide, moisture, &c.—of the fume containing cyanide of potassium deposited in the interior of the pipe employed to collect the gases ; so that the existence of cyanogen *gas* in the lower levels of the blast-furnace is hardly substantiated. *Vide* Lowthian Bell, *Chemical Phenomena of Iron Smelting*, p. 138.

decrease relatively to the nitrogen for several feet upwards from the tuyere, and then *increase* again pretty regularly to the top. Now as any reduction of iron oxide by carbon oxide and subsequent reconversion of the carbon dioxide thus formed into carbon oxide in virtue of the reaction,

$$CO_2 + C = 2CO,$$

must necessarily cause an *increase* in both carbon and oxygen per 100 of nitrogen present, the apparent decrease noticed from the tuyere level upwards for some dozen feet or more is presumably really due to the *evolution of nitrogen gas from the decomposition of the cyanates*, the carbon and oxygen of these salts being mainly converted for the time being into non-gaseous compounds (probably into alkali-metal oxides and carbonates through the further reaction of the cyanates on incompletely reduced oxides of iron), and these compounds being re-converted into cyanides a little higher up, thus simultaneously diminishing the nitrogen and increasing the oxygen in the gases, and therefore as a whole increasing both carbon and oxygen relatively to nitrogen in the upper levels.

35. The pig-iron thus produced in the blast furnace is virtually a solidified solution (§ 102) of amorphous carbon in melted iron, more or less sulphur, silicon, phosphorus, &c., being also present, according to the nature of the ore and fuel used, and other circumstances. One peculiar circumstance connected with this substance is that under certain conditions as to the presence or otherwise of these impurities and more especially as to the time occupied in solidifying, a greater or less amount of the dissolved carbon passes spontaneously into the graphitoidal or black-lead modification which is insoluble in molten iron; when this takes place the pig-

iron solidifies into masses exhibiting a grey colour and well-defined crystalline structure on fracture; from the crystalline surface fragments of actual graphite can be picked out by a penknife point. When this separation and partial crystallization does not take place the fracture is whiter, and the pig more brittle. A good deal of the practical value of pig-iron for the preparation of castings depends on its being what is technically termed "grey," *i.e.*, of such a nature that the graphitoidal separation and partial crystallization can take place readily during the time that the casting takes to solidify : and much of the art of the iron founder consists in blending pigs of various kinds so as to produce a casting of grey-iron where the granulation has taken place to just the extent requisite to give to the casting the maximum strength and power of resistance to strains. "White" pig-iron (in which little or no graphitoidal separation has taken place during the solidification of the pig after running from the furnace) is consequently of less commercial value than grey pig-iron, which again commands a price dependent on its grain (as well as other circumstances) : sometimes a pig will consist of white iron with patches of grey iron interspersed throughout ; in this case it is known as "mottled iron." This spontaneous allotropic alteration of the amorphous or lampblack form of carbon into the crystalline graphitoidal modification whilst dissolved in a solvent, is precisely analogous to a similar phenomenon taking place with amorphous silicon and aluminium ; by dissolving the former in the latter when molten, allowing to cool, and dissolving away the aluminium from the crystalline mass obtained, crystallized allotropic silicon can be obtained. Another phenomenon of the same character is the gradual separation of red phosphorus (insoluble in carbon disulphide) from a solution of the yellow modification in carbon disulphide or certain organic iodides; in these instances the action of light greatly promotes the allotropic change, if indeed it is not essential thereto; in any case, however, the allotropic change of phosphorus is slow compared with that of carbon.[1]

36. Pig-iron which is unsuitable for use in the production of castings in the foundry, or which is not required for such purposes even though suitable, is usually worked up into malleable iron by refining processes, which consist essentially in the partial oxidation of the whole by a current of hot air and the reaction of the oxide of iron thus produced on the oxidizable impurities not yet removed.[2] Certain kinds of pig-iron, however, are treated in other ways, and more especially are directly converted into varieties of steel; thus in Bessemer's process the impurities are almost wholly burnt out by blowing a stream of hot air through the molten mass; in this way, however, it is impossible to produce malleable iron of uniform and good quality, so that to Bessemer's original blowing process is now superadded another step intro-

---

[1] That portion of the amorphous carbon in grey iron which has escaped conversion into graphite is regarded by many chemists as chemically combined with the iron and not merely dissolved therein; the fact that, on dissolving such iron in hydrochloric acid, a fraction of the carbon combines with the liberated hydrogen and escapes as a volatile body being supposed to indicate combination of the carbon with the iron. That this does not necessarily follow, however, is indicated by the circumstance that when hydrogen is liberated in presence of finely divided free sulphur, traces at least of sulphuretted hydrogen, and often much larger quantities, are produced by the direct combination of the nascent hydrogen with the particles of sulphur, which in this case is clearly not combined with any metal.

[2] Thus, in the case of carbon—

$$2xFe + yO_2 = 2Fe_xO_y$$
$$Fe_xO_y + yC = yCO + xFe,$$

and similarly with phosphorus, sulphur, silicon, &c.

duced by Mushet, consisting in the addition to the blown Bessemer product, of a certain amount of fused *spiegeleisen* (a highly carboniferous pig-iron also containing manganese smelted from Spathose ore and deriving its name from the mirror-like crystalline surfaces exhibited on fracture), so as to bring the percentage of carbon in the total mixture up to the amount requisite to form a steely-kind of product, generally designated "Bessemer steel." Only pig-iron free from phosphorus or very nearly so and containing but little sulphur, can be thus successfully treated, as the blowing does not remove phosphorus from the pig-iron employed and only partially removes sulphur, although other noxious impurities, especially silicon, are thus oxidized and burnt out. In Heaton's process the oxidation of the impurities is effected by running the molten pig-iron on to a mass of nitrate of soda in a suitable vessel, the salt being contained in a kind of chamber or cavity at the bottom ; in this way a rapid stream of oxygen is made to bubble through the mass, by which means the same results are produced as by blowing air through in Bessemer's process ; it is stated that sulphur and phosphorous are removed to a greater extent in this way than by air-blowing alone. In Uchatius's process, the pig-iron is fused with a small quantity of iron oxide in a steel melting furnace, so that the oxide is reduced by the carbon of the pig, and a steel results of quality dependent on the amount of residual carbon, &c. In Price and Nicholson's method, and in the Siemens-Martin process, the quantity of carbon in the pig is reduced to that requisite to form a steel by the addition of scrap or other malleable iron nearly or quite free from carbon in suitable quantities : in one of the modifications of the latter process the iron is added as a spongy mass of reduced metal obtained by the action of reducing gases on very pure iron

III.] CHIEF INDUSTRIAL APPLICATIONS. 71

ores; and in another, partially reduced ore in a fused state is added to the fused pig-iron, and spiegeleisen added if necessary to the product to give the proper carbon percentage for steel: in both these modifications Siemens' gas regenerative furnaces are all but indispensable (§ 31).

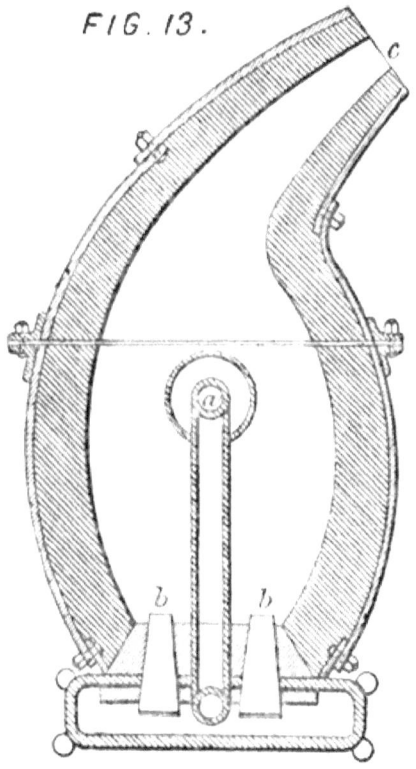

FIG. 13.

37. Figs. 13 and 14 represent two sections at right angles to one another, of a "Bessemer converter," consisting of a pear-shaped iron vessel lined with firebrick, or better, a particular material known as "gannister," and capable of movement about a horizontal axis. One of

the trunnions on which it turns is hollow, and enables a blast of air to be sent into the vessel through the hollow axle, *a*, by a tube to the base of the converter, which is a kind of box, roofed in with a firebrick cover through which pass several tuyeres, *b b b*. By hydraulic power, the converter is turned into an inclined position so that

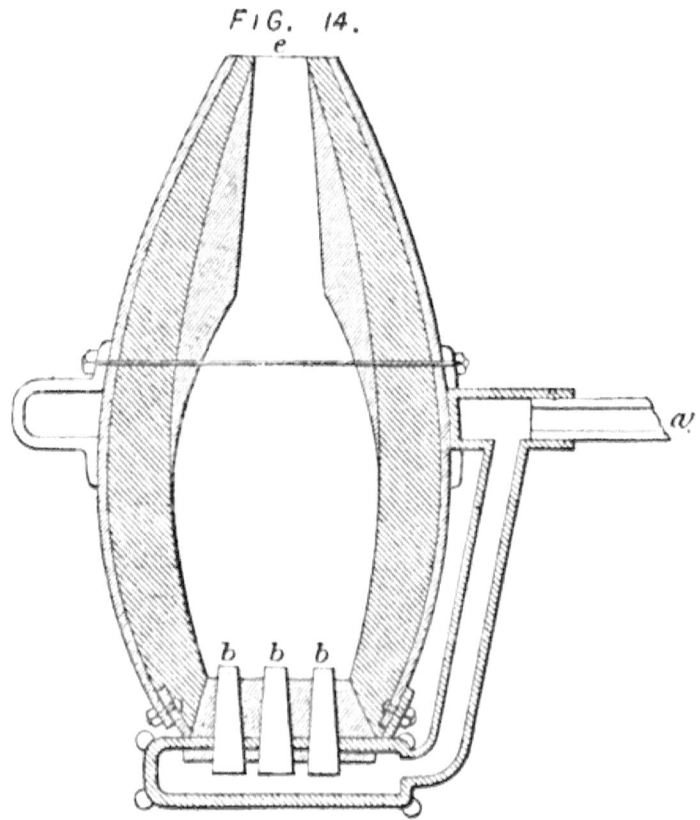

FIG. 14.

the molten metal is run in at the mouth, *c*, the tuyeres being above the level of the fused matter collecting inside; a gentle blast is turned on to prevent the fused metal splashing into the tuyeres and stopping them up by solidifying in them. The blast is then turned on and the vessel brought mouth upwards: when the blowing is completed, the requisite amount of spiegel is added and the blast kept on for a short period to produce perfect intermixture; finally the converter is inverted so that the steel runs out into a large ladle from which it is poured into suitable moulds and formed into ingots or castings. The manganese in the spiegeleisen appears largely to ameliorate the quality of the metal produced, conferring on it a higher power of welding, rolling, and working generally without cracking, and diminishing the tendency towards "red-shortness" (brittleness whilst hot) noticeable in metal containing small quantities of sulphur; hence manganese-iron alloys rich in the former metal are often employed in lieu of spiegeleisens which only contain lesser amounts; these are distinguished as "ferromanganese." This influence of manganese on the physical and useful qualities of steel has long been known in the steel trade, the addition of manganese carbide to the materials used in steel-making having been patented by Heath in 1839, and very largely adopted since in one form or other.

38. Previously to the introduction of the Bessemer process, the product of which has, to a very considerable extent, superseded malleable iron for many purposes, the production of workable soft iron from pig-iron was effected by two processes still largely employed; in one of these, known as the "charcoal finery," the pig is melted in a small furnace and covered with a heap of charcoal (coke is frequently employed, the expense of charcoal preventing its use for any but the finest qualities, such as those employed for tin-plate making): powerful jets of air are then directed on to it from tuyeres for some time; in this way the silicon, sulphur, carbon, and phosphorus

present are oxidized, and a cinder formed mainly consisting of silicate and phosphate of iron; considerable loss of metal usually attends this mode of operating.

The process of "puddling" is in principle the same as that of the refinery, save that the fuel is usually coal, and

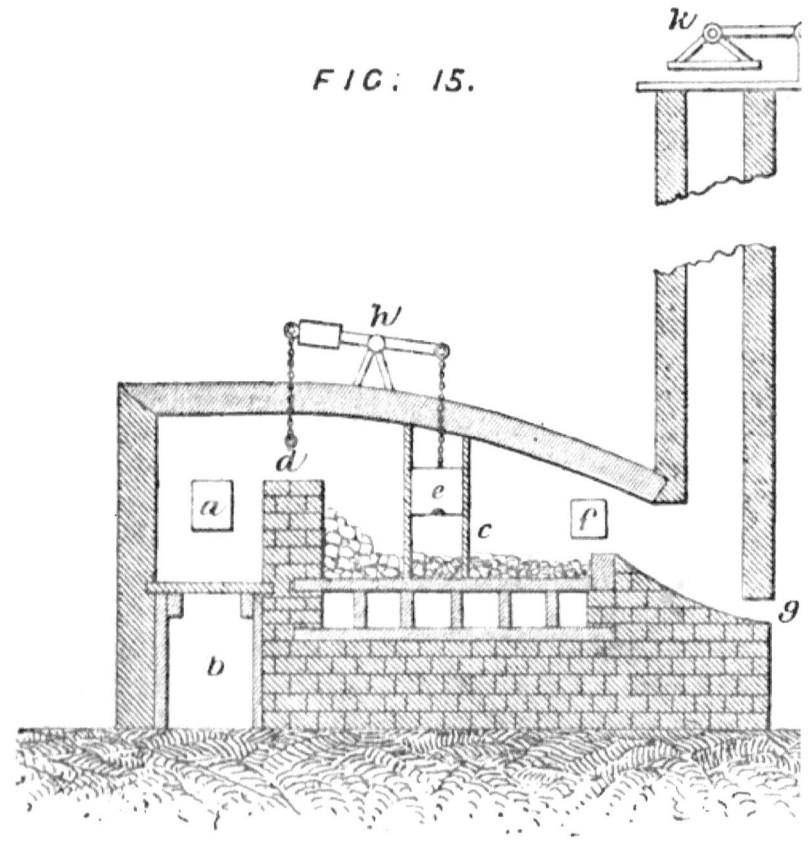

FIG. 15.

is burnt in a separate fireplace; the flame is made to play over the iron to be puddled by means of an arched roof, Fig. 15 (whence the term reverberatory furnace). a is the door through which fuel is introduced, b, the ashpit. The hearth, or sole of the furnace, *c*, is separated from the fireplace by a wall of brickwork, *d*, termed the "bridge," over which the flame plays on to the metal on the hearth. Through a working door, *e*, the puddler stirs up from time to time the charge with an iron *rabble*, so as to expose it to the air, and thus oxidize it, and to incorporate with it a certain amount of oxide of iron used as a covering for the hearth and termed the *fettling*. The hearth itself rests on iron plates, supported on iron girders, with air spaces underneath to diminish the heating action on the plates, and the whole furnace is cased externally with iron, and strongly bound together; at the end furthest from the fire is a charging door, *f*, through which fresh pigs are introduced, and at the extreme end, at *g*, a "taphole" or "cinderhole" is fixed for running off the fluid *tap-cinder* (silicate of iron, &c.), formed during the process. The draught of the fire, and hence the temperature, is regulated by a *damper*, *k*, on the top of the chimney, opened and closed by the workman by means of a chain or rod attached to the lever; the doorhole is provided with a door supported by a chain and counterpoise, so that when the charge is to be removed, the door is lifted and a large space then left; whilst rabbling the charge the entrance of cold air is avoided by means of a small opening or notch at the bottom of the door through which the shank of the rabble is passed: when not thus required, this opening is closed by a brick, or a shovelful of cinders, &c. As the silicon and carbon are removed from the pig-iron, the substance, at first liquid, becomes pasty, and finally spongy masses of iron separate from the fluid cinder; these are raked and worked into one mass termed a "ball" or "bloom" when the metal has wholly "come to nature"; this is extracted from the furnace through the door-hole, and quickly transferred to powerful jaws or squeezers, or hammered under a steam hammer (*shingled*) until it forms a coherent mass; during the shingling, silicate of iron and fused oxide of iron are squirted out on all sides by the blows, constituting "slag" (*schlagen* = to strike). The puddled ball is usually further worked by forming into bars under the hammer, cutting them up, welding together the fragments (after "piling" or "faggoting" them into bundles and heating to a welding heat in a "reheating furnace") and again rolling into bars, and in some cases repeating the process; this develops a fibrous structure in the bars or plates produced, greatly adding to their strength. Sometimes the operations of refining by the finery and puddling are combined, the metal being partially purified by the first process, and then finished off by the second; indeed, by some the term "puddling" is restricted to the application of the process above described to the refined or partially purified metal, the conversion of pig iron into malleable iron in one operation being designated as "pig boiling" on account of the ebullition of the fused pig caused by the rapid escape of carbon oxide when the action is at its height: others use the term "puddling" as a generic name for both operations, and call the puddling of refined iron " dry puddling " on account of the smaller quantity of fluid cinder formed. Sometimes the "pig-boiling" operation is applied directly to the liquid metal tapped from the blast furnace without casting into pigs and cooling; in this way fuel is saved, but many practical inconveniences are introduced.

The "fettling" of the puddling furnace becomes to a considerable extent reduced to the metallic state, and thus helps to balance the loss of metal caused by oxidation and conversion into fluid silicate of iron, &c. (*tap-*

cinder). By subjecting this silicate to a kind of liquation process, part runs away as a more fusible cinder, and part remains as a difficultly fusible mass: this latter is often used as a covering for the hearth of the puddling furnace under the name of "bulldog."

39. To avoid as much as possible the enormous amount of manual labour of a most exhausting kind entailed in these processes, machines have been constructed to effect automatically the same kind of operations: in one of these, known as "Danks' rotary puddler" (Fig. 16), the mechanical agitation is effected by placing the pigs inside a kind of iron drum, c, lined with firebrick and supported on friction rollers, so that the drum itself forms the reverberatory furnace, being applied to the flue end of the fireplace, the flame from which is regulated by two air blasts, one above the fire bars, a, the other below, b; a large cog-wheel surrounds the drum so that it can be made to rotate at any required speed; a movable hood, d, leading to the chimney is applied to the far end, supported by a crane, so that when d is thrust on one side, the end of the drum is open, serving as a charging and withdrawing door. The action of the machine in producing mechanical agitation of the charge is greatly enhanced by the peculiar form of lining employed; some tolerably fusible iron ore is placed inside and melted by the flame; the drum being brought to rest, a pool of liquid matter is formed at the bottom; lumps of a very hard and infusible ore are thrown into this pool, whereby the fluid matter is chilled and solidified, thus cementing the lumps to the inside. The same process is repeated, the pool being now formed in another place, and so on, until the whole of the inside of the drum is lined with irregularly projecting infusible lumps. Several modifications of the machine have been brought out

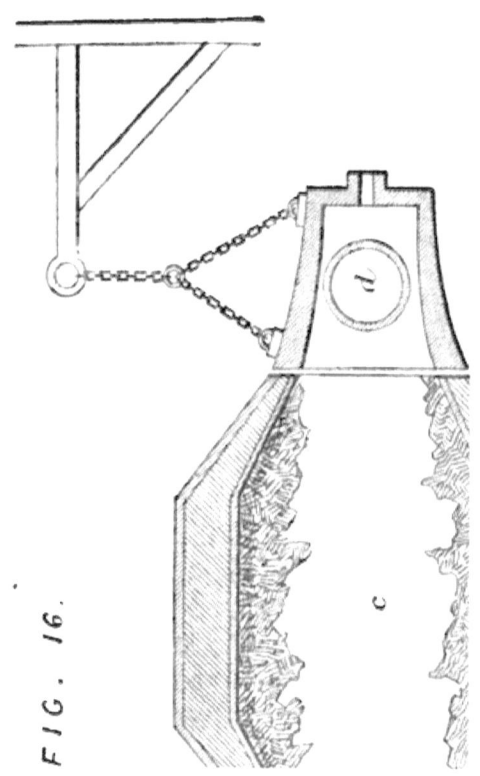

FIG. 16.

subsequently by other inventors: puddling by machinery, though successful up to a certain point, can as yet, however, hardly be said to have advanced far beyond the experimental stage, saving perhaps for some special kinds of material.

40. The steels formed by most of the processes above alluded to ("direct processes" and methods based on the partial decarburization of pig-iron, §§ 31, 36), though serviceable for many purposes, are not, as a rule, well adapted to the manufacture of tools and cutting instruments generally, although steel of the highest qualities can be formed by some of these methods by paying special care and attention, and choosing suitable substances to start with. A steelmaking process, known as *cementation*, and based on entirely different principles (viz., the carburization of malleable iron), is largely employed for the production of the finer kinds of steel; this process is of much greater antiquity than most of the other methods above described; it consists in packing bars of the purest malleable iron in boxes alternately with layers of charcoal (nitrogenous organic matters, such as horn-shavings, ferrocyanide of potassium &c., being often also added), and then heating the whole to a high temperature for a lengthened period (eight to twelve days). The bars of soft iron thus become carburized and converted into "blister steel" (so called from the occurrence of blisters on the bars). Probably the rationale of this conversion is the absorption of carbon oxide by the metal, and its reduction, with separation of carbon and oxidation of the metal, thus—

$$x Fe + y CO = y C + Fe_x O_y,$$

the iron oxide thus produced being again reduced to the

metallic state by another portion of carbon oxide, thus—

$$Fe_xO_y + yCO = yCO_2 + xFe.$$

That these two actions can go on simultaneously under certain conditions has been shown by Lowthian Bell (*Chemical Phenomena of Iron Smelting*), nickel and cobalt (but apparently no other metals) possessing the same power.

The blister steel thus produced is not uniform in character, the outsides of the bars being, as might be anticipated, more highly carburized than the insides; to obtain a homogeneous steel, the blister steel is melted and cast into ingots of "cast steel"; "shear steel" or "tilt steel" is made by cutting up bars of blister steel, binding in bundles, heating to a welding temperature, and then hammering under a tilt hammer. For cutting instruments the crystalline character possessed by cast steel is objectionable, although the uniformity of structure produced by fusion is indispensable; hence the cast steel is heated and carefully forged and rolled so as to develop a fibrous character.

41. The great value of steel arises from the peculiar effects produced on it by rapid cooling after heating: it thus becomes excessively hard and considerably brittle; by gently heating hardened steel and allowing it to cool slowly this brittleness is much diminished or altogether removed, and great elasticity communicated instead; the degree of "temper" thus given to the steel by this *annealing* or *tempering* depends on the temperature to which the hardened steel was heated and from which it cooled down slowly; the higher the temperature, the softer the steel: in practice, this temperature is judged of by the colour exhibited by the surface of the steel

(previously polished) on heating: thus the following colours correspond to the temperatures annexed:

Colour.	Temperature.
	° °
Pale yellow to dark straw	215—260
Brown yellow shading into purple	260—275
Purple to dark blue	280—300
Pale blue passing into green	305—340

Steels tempered at pale yellows are preferred for most tools for metal working; darker straw shades for screw cutting and tools for working in wood; brown yellow to purple shades for instruments that have to resist blows or vibration, such as chisels, hatchets, saws, &c.; dark blue and purple shades are only employed when great elasticity is required, as for clock, watch, and other springs; whilst tempering at pale blue to green renders the steel so soft as to be quite unable to bear a cutting edge.

As soft steels can be readily welded to malleable iron, many tools are made chiefly of soft iron with a strip of steel affixed where the cutting edge or face is placed: thus in the production of dinner-knives the body of the blade is of soft iron with a comparatively thin steel face. Frequently heavy iron goods are partially steeled on the surface (*case hardened*) by placing them in contact with some carbonaceous matter, especially ferrocyanide of potassium, keeping at a high temperature, and finally plunging into cold water to harden: a film of steel of $\frac{1}{20}$ to $\frac{3}{8}$ inch in thickness can thus be produced.

In order to procure steels of great hardness for special purposes, various combinations of iron with minute quantities of other metals have been proposed, but the practical value of many of these is as yet rather doubtful. Thus, according to Faraday and Stodart, rhodium, silver, or chromium improves the quality when added in minute quantity to steel. Titanium steel and tungsten steel have also been prepared and manufactured; the latter is extremely hard; as yet, however, none of these substances have come into anything like extensive use.

42. **Copper.**—The principal sources of this metal are the Australian ores, consisting chiefly of carbonate; those from North America and Siberia, mainly composed of native metal; and the various forms of copper pyrites or double sulphide of iron and copper, and copper glance largely mined (together with other cupriferous minerals), in Spain, Chili, Cuba, Saxony, Russia, Wales, and more especially Cornwall. The pyrites of the Huelva and Tharsis Mines in Spain and Portugal contains only small quantities of copper, and is ordinarily treated by a wet process (§ 43) after the sulphur present has been utilized for the manufacture of oil of vitriol: certain ores, consisting of copper carbonate disseminated through sandstone and other rocks, are occasionally worked by another more simple wet process consisting of the solution of the copper by treatment with hydrochloric or sulphuric acid, and its precipitation from the solution by scrap-iron.[1] In a similar fashion copper is extracted from the pump-waters of the copper-mines by simply passing them into reservoirs or tanks containing scrap-iron; the copper gets into solution in the water from the spontaneous oxidation of the copper pyrites forming soluble copper and iron sulphates. The chief method of extracting copper from its ores however is that known as the " Swansea process," from its being carried out on a vast

---

[1] $CuCO_3 + 2HCl = CO_2 + H_2O + CuCl_2$
$CuCl_2 + Fe = Cu + FeCl_2$.

CHIEF INDUSTRIAL APPLICATIONS.

scale at that locality, something like 20,000 tons of copper being annually smelted there. The principle of the process is simple enough, it being based on the circumstances that if copper pyrites be exposed to the action of hot air, the iron sulphide is oxidized first, so that copper sulphide free from iron is first obtained; and that when this is further oxidized, the copper oxide formed reacts on the yet unoxidized sulphide, forming sulphur dioxide, which escapes, and metallic copper: the practical carrying out of these operations, however, is sufficiently complex, at least six stages being recognisable: these consist of (*a*) preliminary calcination of ore, whereby much of the iron sulphide is converted into oxide whilst arsenic and sulphur are expelled as oxides: (*b*) first fusion of calcined ore; in this process lime, sand, fluorspar or other fluxes are added when requisite, so that the earthy matters of the ore and the majority of the iron oxide fuse together to a cinder, whilst impure cuprous sulphide sinks to the bottom, whence it is drawn off through a tap-hole in the furnace and run into water. By judiciously mixing together suitable ores, such as carbonate and roasted copper pyrites, &c., with suitable admixtures of ores containing siliceous gangue, or of siliceous cinder from later operations, a tolerably uniform product is obtained, designated "first *matt*," or "*coarse metal*." The object of running the fused matt into water is to disintegrate it and render it ready for the third operation (*c*), which consists of roasting the granulated coarse metal for several hours, so as to remove much sulphur and oxidize most of the remaining iron. (*d*) melting calcined coarse metal; the roasted coarse metal is mixed with suitable quantities of a rich copper ore containing silica with but little iron, or certain qualities of cinder from the subsequent operations, and the whole fused so as to form another siliceous iron cinder, and a nearly pure cuprous sulphide or "fine metal" (second matt): the cinder thus produced generally contains copper which is extracted by using it for admixture with other substances in the production of coarse metal or by other special treatment. The fine metal thus produced is cast into pigs, and consists of nearly pure cuprous sulphide, with a small admixture of iron sulphide, so that the term "metal," as applied to it and to the first matt, is, strictly speaking, incorrect. In the fifth operation (*e*) actual metallic copper is produced, the pigs of "fine metal" being roasted in a reverberatory furnace (§ 38), so as to burn off much of the sulphur and form copper oxide ; on raising the temperature so as to fuse the whole, the molten mass appears to "boil," from the rapid evolution of sulphur dioxide, metallic copper being set free ; ultimately this is drawn off by a tapping hole; the scoriæ from this operation are generally highly cupriferous and are worked in with the mixture employed in the fourth stage (*d*). The copper thus produced is run into moulds and is then designated "blister copper," as it is full of cavities from the gases with which it is charged ; it generally retains small quantities of foreign substances, such as iron, sulphur, arsenic, &c. To remove these the sixth operation (*f*) of refining is resorted to ; this consists in roasting the ingots of blister copper so as to oxidize them superficially, the porous character of the metal allowing also of a considerable amount of internal oxidation ; finally, the heat is raised so as to fuse the whole, when the oxide of copper previously formed oxidizes most of the impurities, producing a much purer metal : the scoriæ are raked off and the oxide of copper disseminated through the molten metal removed by placing powdered charcoal or anthracite on its surface, and stirring up the whole with a pole of green wood, usually birch: if the metal be "over-poled" it becomes somewhat brittle, probably from its taking up a little carbon: this is removed by skimming off the anthracite and exposing the fused copper to the air a while. When the poling operation has been carried on to the right extent (known by continually sampling the metal by means of a ladle and examining the physical characters of the small ingot produced by quickly cooling the ladle and contents in water), the copper is cast into ingots, or is granulated by pouring into water, forming, according as the water is cold or hot, *feather-shot copper* or *bean shot*, the latter being in rounded granules, the former in less regular shapes.

Sometimes operations *c* and *d* are again repeated in order to obtain a yet better "fine metal;" not unfrequently a little lead is added in the refining operation. This metal oxidizes in the hot air-current, and the oxide so produced facilitates the removal of arsenic, iron, antimony, &c., by yielding oxygen to these impurities: finally the lead is itself removed by continually stirring the fused metal and raking off the scoriæ before poling. In some cases, the cinders of the various operations are not directly intermixed with fresh materials for one or other of the above-mentioned stages, but are subjected to special fusion operations with a view to obtaining from them the copper present intermixed with lesser quantities of the valueless silicates of iron, &c., constituting the majority of the cinder: these special treatments of scoriæ, when adopted, are usually applied to the cinder from operations *b* and *d;* that from the former often contains metallic grains of copper diffused through the mass, which are united together by simple careful fusion; that from the latter process is fused with coal or other reducing agents, when the copper present is reduced along with portions of other metals, forming a white brittle alloy which is worked up with a fresh charge of coarse metal in the fourth operation (*d*).

43. The wet process by which poor cupriferous pyrites is worked for copper is commonly known as "Henderson's process," from the name of the patentee. This consists in roasting the "burnt ore" of the vitriol-maker (pyrites that has been burnt in specially-constructed

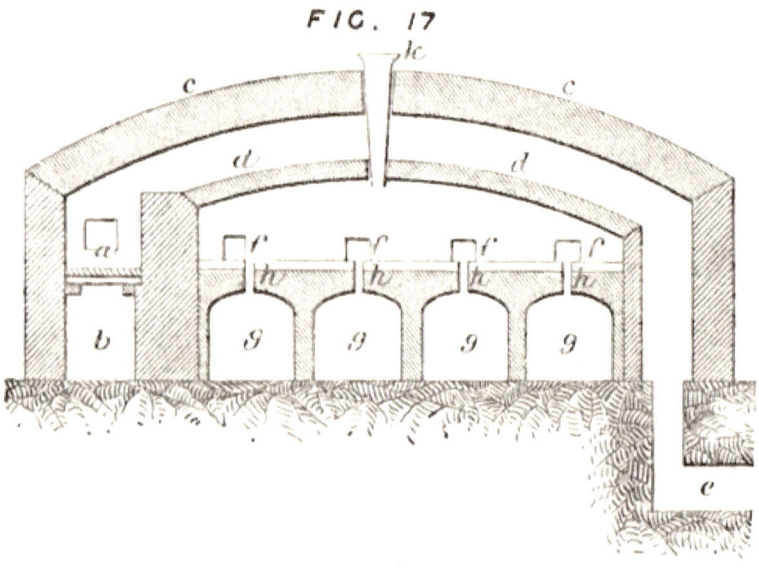

FIG. 17

roasters or "kilns," so as to utilize the sulphur dioxide produced) with common salt in a peculiarly constructed kind of furnace where the flame does not play directly on the materials to be heated, but passes from the fire over a thin brick-arched roof surmounting the bed or "sole" of the furnace, so that the substances to be roasted are heated by radiation from the nearly red-hot brick arch; in fact resembling in construction the "salt-cake furnace" of the alkali-maker, saving that, as a somewhat less heat is required, the flame is not made to pass through flues under the sole of the furnace after heating the radiating brick arch. Fig. 17 gives a general idea of the character of furnace employed; the flame from the fire, a, passes between the external roof of the furnace cc, and the thin inner arched roof dd, and thence to the chimney through the underground flue e (practically, the waste heat is utilized before the gases pass to the chimney by making them heat evaporating pans, &c.); b is the ashpit, and $ffff$ working doors for stirring the charge from time to time as the operation progresses. The bed of the furnace is supported over a series of arches (something like a railway viaduct); in the centre of each arched vault thus formed is a cavity, $hhhh$, left in the brickwork of the sole, closed by a tile, so that when the charge is sufficiently roasted the substance is withdrawn from the furnace through these perforations into "bogies," or trucks, placed underneath to receive it, fresh materials being added through the hopper, k. The burnt ore usually contains about 3 to 3·5 per cent. of sulphur and 3 to 4 of copper; if too little sulphur be present relatively to the copper it is mixed with a suitable quantity of the dust of unburnt cupriferous pyrites, so as to make the sulphur and copper stand to one another in about the ratio of three to two: about four parts of salt for one of copper are then added, and the whole roasted in the furnace for about six hours, the heat being not allowed to become too great, otherwise the copper chloride formed is further decomposed and rendered less readily soluble in water. Fumes of hydrochloric acid are evolved and condensed by a tower filled with coke kept moist by water trickling down exactly as in the salt-cake process; a certain amount of the chlorides of copper, iron, and arsenic is also thus condensed. The acid liquor thus produced is used to wash out the last portions of copper from the residue left on treating with water (*lixiviating*) the mixture of chlorides of sodium, copper, and iron, sulphate of soda, and ferric oxide, &c., formed by the roasting in the furnace. Usually the pyrites contains minute quantities of silver which become transformed into chloride, and this, though almost insoluble in water, is sufficiently soluble in brine to be dissolved out by the aqueous solution obtained, a considerable amount of the salt originally added remaining unchanged and thus furnishing the solvent. From the first aqueous liquors the silver is separated as described in § 21 ; from the residual liquors left after precipitation of the silver the copper is thrown down by means of scrap-iron. Usually the liquors obtained by washing the insoluble residue left after the first lixiviation with water with the acid watery solution from the tower furnish a less pure copper on treatment with iron, and therefore are worked up apart. When the roasting has been properly conducted, the substance finally left undissolved is all but free from copper, and consists mainly of ferric oxide with more or less quartzose matter. This "purple ore" is sometimes of sufficiently good colour to be used as a pigment; it furnishes a good material for "fettling" puddling furnaces (§ 38), and is sometimes used up in the copper extraction works as a source of metallic iron to precipitate the copper, the oxide being dried and reduced either by heating with small coal, or by the gases produced in an arrangement somewhat analogous to a Siemens' gas-producer (§ 31): the copper precipitated by the reduced purple ore is somewhat contaminated with unreduced iron oxide, coke dust, quartz, &c., but these impurities are readily removed during the melting and refining processes through which the copper is put before it is suitable for manufacturing purposes.

The theory of this wet process for copper extraction is that under the oxidizing influence of the hot air in the roaster the copper sulphide and common salt become transformed into copper chloride and sodium sulphate, thus:

$$Cu_2S + S + 4NaCl + 4O_2 = 2CuCl_2 + 2Na_2SO_4,$$

the extra sulphur (assuming the copper to be present as cuprous sulphide) being derived from iron sulphide also present, the iron of which becomes converted partially into ferric oxide and partially into ferric chloride by a somewhat analogous action.

44. Many other processes for the extraction of copper are in use in various localities; for the most part these are modifications of the Swansea method. At Mansfeld, in Saxony, an alloy, containing about 95 per cent. of copper and 3 of iron, with small quantities of silver, sulphur, and other impurities, is smelted in a small blast-furnace, by means of charcoal and coke, from the well-calcined ore of the locality. This is an argillaceous schist, throughout which copper pyrites is disseminated to the extent of a few per cent. only. From the alloy thus obtained the silver is separated by liquation with lead (§ 20); or the silver is removed from the calcined ore before its treatment in the blast-furnace by heating the ore with salt, and then putting it through the Freiberg amalgamation process (§ 23). Finally the impure copper left after liquation, and that directly smelted after removal of silver, are refined by roasting in air and fusion, &c., as in the Swansea process. Similar modes of procedure are employed at some works in Sweden and Norway.

45. **Lead.**—By far the most common ore of this metal is *galena*, or lead sulphide; the sulphate also occurs in moderate quantity in Australia and the United States, whilst the oxide, chloride, carbonate, phosphate, and chromate, though not exactly rare minerals, are still so far scarce as not to constitute any very considerable sources of the metal. Galena almost invariably contains silver in quantity varying with the locality, from less than an ounce to several hundred ounces per ton of metal; usually gold is also present to a much less extent.

As a first step the galena is separated as far as possible from siliceous gangue by crushing and washing with water so as to wash away the lighter earthy particles; the ore is then roasted on the bed of a reverberatory furnace, usually with the addition of a little lime to serve as a flux for the residual gangue, &c., the whole mass being stirred up with a rabble from time to time so as to facilitate oxidation. In this way a mixture of lead oxide and sulphate is formed. When this change has progressed sufficiently far the temperature is raised by regulating the dampers; the mass fuses and boils somewhat as in the analogous operation of smelting copper by the Swansea process (§ 42), and from the same cause, viz., the evolution of sulphur dioxide and formation of metal (*vide* § 9, foot-note). The cinder formed in this operation often retains lead, which is separated by a further operation, somewhat as the analogous copper scoriæ from the first melting in the Swansea method.

46. In some places the calcination of the ore and the subsequent fusion and reduction are made into two distinct operations carried out in separate furnaces. Sometimes the complete separation of lead in the second stage is facilitated by the addition of pieces of metallic iron, which, combining with the sulphur, liberates metallic

lead. In what is termed the "Scotch hearth" the reduction is effected in a forge supplied with a blast of air from a tuyere, peat being the fuel preferred. The calcined and semi-fused ore, technically termed "browse" or "brouze," is placed on the glowing peats, and a little lime added; ultimately most of the metal is separated and collects in a cavity at the side of the furnace, flowing out from the hearth when the fused metal rises up to a certain level; the remainder forms with the earthy matter a "grey slag," which is usually smelted by a separate process to recover the residual lead. At Freiberg a blast-furnace some 20 to 25 feet in height is employed for the smelting of an ore which contains silver and copper.

When much siliceous matter is present in the lead ore, the slags are apt to contain much silicate of lead. To avoid the loss of lead thus rendered imminent cast-iron is mixed with the ore and the whole smelted in a blast furnace. Whatever mode of smelting is adopted, arrangements are employed for intercepting and condensing the fumes of lead carried away from the furnace by the escaping gases: either a series of condensing chambers is provided through which the gases pass before entering the chimney, or, where circumstances permit, a long gallery or tunnel is employed as the chimney, running up the side of a hill for a long distance, in some cases several miles. The "fume" which collects in these flues and chambers is periodically collected and smelted to obtain the contained metal.

Frequently the lead smelted from galena is somewhat impure, containing tin, antimony, copper, &c. Lead prepared from a given ore in the Scotch hearth is generally more pure than that otherwise smelted, because a perceptibly lower temperature is employed in the reducing operation, and hence foreign metals are reduced to a lesser extent. In order to remove silver, the processes of Pattinson or of Parkes (§ 20) are employed. To soften and refine lead containing foreign metals the lead is melted in a pot and then run on to the hearth of a reverberatory furnace, or it is fused in the reverberatory itself. The contained metals and some lead then oxidize and form scoriæ, which are removed by skimming. When the scoriæ consist of little but lead oxide (which is known by their appearance, or judged by taking a sample of the melted lead, cooling it, and examining its physical character and appearance when solid), the purified lead is run into moulds, forming *pig-lead*. With tolerably pure lead this refining operation is complete in a few hours, whilst with impure leads, such as those from Spanish ores, it is not complete till after two or three weeks heating and oxidation.

47. **Tin.**—There is good evidence that this metal was known about 3,000 years ago, it being apparently referred to in the Mosiac books. Since the term "Kassiteros" is, according to Humboldt, derived from the Sanscrit "Kastera," and the word *tin* (*tenn*, Swedish; *den*, Icelandic; *zinn*, German; *étain*, French; *stannum*, Latin) from the Malay and Japanese *timah*, it is probable that the earliest sources of this metal were rather the Indian islands than the south-west coast of England, although it is tolerably certain that the Phœnicians traded with Cornwall at a period antecedent to the foundation of Rome. At present the chief sources of the ores of this metal are Cornwall, Banca, and Malacca; some more or less considerable amounts being also derived from Bohemia, Saxony, Austria, Australia, the United States, Mexico, South America, and a few other places. The chief ores are the oxide (*cassiterite*, or tin-stone) and an

impure sulphide (tin pyrites, or *stannine*), of which the former is much the more important. Frequently it is met with in the beds of streams (whence the term *stream-tin*), as rounded masses of tin-stone washed out from the veins higher up. This form of ore yields a very pure metal, and is consequently sought after with eagerness, not only in streams but in alluvial and estuarian deposits.

Mine-tin, or ores brought up from the mines, is of varying character. Sometimes it is mixed with large amounts of wolfram, arsenical pyrites or copper pyrites, when special modes of treatment are necessary. Generally a certain amount of foreign metallic ores and of gangue is contained. These are separated, firstly, by stamping and washing the ores so as to separate by gravity the heavy tin-stone particles from those of lighter minerals and stones; and, secondly, by calcining or roasting in a reverberatory furnace with continual stirring, or in an automatic calciner, such as that used for the production of arsenic (§ 59). By this means sulphur and arsenic are expelled, iron is oxidized, and copper is partially converted into sulphate. After this roasting the ore is weathered, or exposed to the open air, for some time, so as to complete the oxidation of the copper pyrites. The sulphate of copper thus formed is dissolved out by water, and the copper thence obtained by precipitation with metallic iron (§ 42). The oxide of iron is mostly removed by washing, being comparatively light. The roasted and washed ore is then heated with "culm" (a kind of anthracite), and a little of some flux, such as fluor-spar or lime, in a reverberatory furnace, care being taken that the temperature is not too high, otherwise much tin would be lost in the slag. The metal readily subsides in a fused state, and is then drawn off through a tapping hole into an iron vessel in which it is allowed to stand (whilst still kept fused) for some time, so that all scoriæ may rise to the surface.[1] Finally the metal is cast into ingots, whilst the slags and scoriæ are generally worked over again so as to recover any globules of metal retained, &c. If the tin is fairly pure, the ingots will be capable of breaking up into peculiarly-shaped elongated splinters (known as "grain tin"), when they are thrown down on a hard floor from some little height whilst heated nearly to the melting-point. Accordingly tin is often sent into the market in this form, as a sort of guarantee of its purity. Impure metal requires refining: this is effected by a liquation process, the ingots being gently heated in a reverberatory furnace ; tolerably pure tin flows away first, copper, iron, &c., remaining as an unfused mass, known as "hardhead"; the last portions thus running away are so far impure as to require refining over again. The purer tin thus sweated out is then stirred up in a melting-pot with wet wooden poles, whereby some of the impurities (as well as some of the tin itself) are oxidized and brought to the surface as a scum, which is carefully removed and utilized as an impure tin ore. After standing some time in a fused state, nearly perfectly pure tin rises to the top, whilst the heavier portion at the bottom contains most of the residual impurities. The top portion is carefully ladled off and sold as "refined tin," and the bottom portion either liquated or refined over again, or sent into the market as "block tin" in ingots. This is usually too much contaminated with other metals to be capable of forming good "*grains*" by falling on a hard surface whilst hot.

[1] The reduction is brought about by the conjoint action of the carbon of the intermixed culm and the carbon oxide of the gases in the furnace—

$$SnO_2 + C = CO_2$$
$$SnO_2 + 2CO = 2CO_2.$$

Formerly tin was often smelted by means of a small blast-furnace with charcoal, and this process is still used on the Continent; but it is considered by English smelters that this method furnishes a smaller yield and is more costly than that above described; hence it has been given up in this country. A machine for performing automatically the calcining and stirring operations in the preliminary treatment of the mine-tin is sometimes adopted, and a little salt added to facilitate the expulsion of arsenic and sulphur and the conversion of copper pyrites into soluble copper compounds (as in Henderson's process for the extraction of copper from low percentaged cupriferous pyrites, § 43).

48. **Zinc.**—The peculiar properties of the alloy of this metal with copper, termed *brass*, have been known from a very early period, many of the so-called antique bronzes containing so much more zinc than tin as to approach much more nearly to brasses than to bronzes in character and composition: these alloys were probably obtained by smelting together a natural or artificial mixture of minerals containing copper and zinc, the processes for the isolation of the latter metal being of comparatively modern origin, depending on the volatility of the metal at a red-heat, *i.e.* being processes of distillation. The chief ores of zinc worked for metallurgical purposes come from New Jersey and the United States, Belgium, Silesia, and Spain, considerable deposits being also found in various parts of Great Britain. They consist mainly of oxide, sulphide, carbonate and silicate, the former being found in New Jersey: *zinc blende* (the sulphide) usually accompanies galena and sometimes pyrites, whilst *calamine* (the carbonate) is found in Derbyshire without intermixture with lead ores, though often it occurs in the vicinity of galena: the

silicate (*electric calamine*) is chiefly imported into this country.

When blende (known as "black jack" from its colour when tolerably pure save a little iron sulphide) is employed as a source of zinc it is subjected to a prolonged roasting for the purpose of burning off the sulphur and forming an oxide of zinc;[1] if lead be contained this roasting must be carried on for a longer period than if the blende be free from this metal: from the oxide thus produced the metallic zinc is reduced by mixing with small coal (anthracite) and heating in a distilling arrangement, in principle not unlike the "capellina" apparatus used for silver amalgam (§ 22), or in fireclay retorts like those used in the gasworks. Calamine is usually roasted before being submitted to the distillation process, for the purpose of expelling moisture and carbon dioxide and of opening the pores of the mass and rendering it easy to pulverize and mix with the powdered anthracite; frequently a mixture of calcined blende and calamine, or of red zinc ore (oxide of zinc) and calamine is employed instead of one kind of ore only.

49. The character of the distilling arrangement employed varies considerably in different localities; one of the oldest forms (Fig. 18) consist of a number of large crucibles filled with the mixture of zinc ore and small coal and heated by a reverberatory furnace; each crucible is closed with a cover luted on air-tight: the vapour of zinc and the carbon oxide formed by the heating are conducted out of the crucible through a hole in the bottom, into which is cemented a fireclay pipe passing down through the floor of the reverberatory furnace into a vault below; iron tubes are affixed to the lower ends of

[1] $2ZnS + 3O_2 = 2ZnO + 2SO_2$.

these fireclay pipes, dipping downwards into vessels containing water, so that the zinc condenses in the tubes and runs down into the vessels underneath. To prevent the mass in the crucibles falling down the tubes blocks of wood are inserted into the fireclay pipes ; these become carbonized by the heat, and are then sufficiently porous to allow of the vapours passing downwards through them, whilst retaining sufficient strength to prevent the superincumbent solid matter from breaking through them. The iron tubes are cleared out by a rod from time to time lest the condensed zinc and the "fume," or oxide, carried over should block them up.

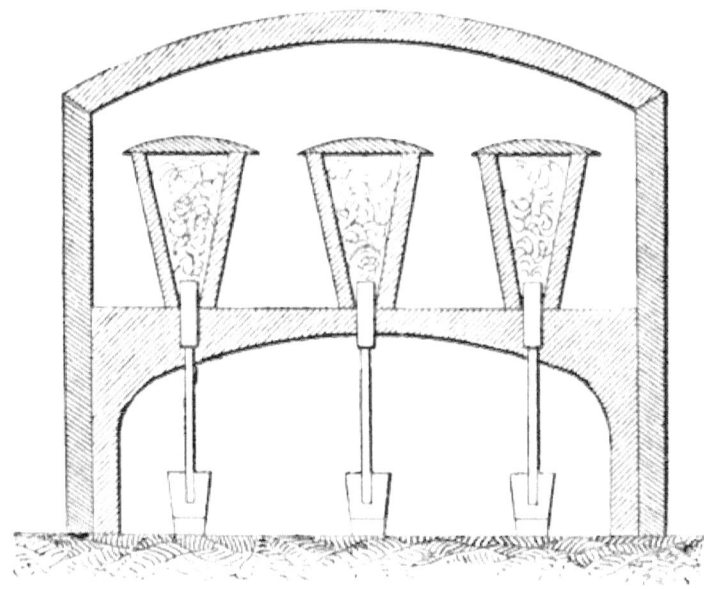

In Silesia retorts are used shaped like small gas retorts or large muffles; the vapours are led away into a rectangular downward condensing tube, luted on to

H

the mouth of the retort; several of these retorts, sometimes twenty, are mounted in a "bench" and heated by the same furnace. In what is termed the Belgian process, a large number of retorts are employed mounted in a kind of kiln, each retort being placed at an angle of about 30° with the horizon, the mouths being lowest: short clay condensing pipes are luted into the mouths of the retorts: to these are affixed conical wrought-iron pipes with a narrow terminal orifice some $\frac{3}{4}$ inch in diameter: the metallic zinc collects in the clay pipes, whilst the iron conical terminals retain a considerable amount of zinc oxide which is collected and used over again with a fresh charge. As many as eighty retorts are sometimes mounted in the same bench, to equalize the rate of working off; those at the top are usually made smaller and are charged with more easily reducible ores, so that the lesser heat communicated to them may not interfere with the process, as a whole, by causing delay through great differences in the time required to work off the charge in each retort. The waste heat from the bench of retorts is employed to roast and calcine ores for a new operation, and to dry and season the fireclay retorts and tubes kept in stock for renewals in case of breakages, &c.

50. When the zinc ores used contain *cadmium*, this metal is chiefly contained in the first portions of zinc that distil over; the fume which is condensed at the same time is brownish and contains much cadmium oxide; from this fume, or from the cadmiferous zinc, metallic cadmium can be separated, best by a wet process, consisting of solution in acid, precipitation of cadmium sulphide by a current of sulphuretted hydrogen, whereby this metal is completely separated from zinc, conversion of the sulphide into carbonate by solution in acid and

precipitation by an alkaline carbonate, and finally distillation of the carbonate (or of the oxide prepared therefrom by gentle ignition) with lamp-black, whereby metallic cadmium is obtained as a distillate. Metallic cadmium, however, is but little used, the main industrial application of this element being the use of its sulphide as a brilliant yellow pigment : certain of its salts (iodide and bromide) are also employed in medicine and for photography.

CHAPTER IV.

METALLURGY OF THE LESS IMPORTANT OXIDIZABLE METALS.

51. OF the remaining metals, few are of great industrial importance in the reguline state; aluminium and, in a less degree, magnesium are used to some extent, whilst the former of these, as well as nickel, antimony, bismuth, arsenic, and manganese, are of value for the preparation of sundry important alloys.

Aluminium.—The metallurgy of aluminium differs from that of any other of the above described metals, inasmuch as it is impossible to obtain the free metal from any of its natural sources, abundant though these are, by the ordinary methods of smelting, &c.; and processes for converting these natural sources into compounds that can be economically dealt with are a necessary first step towards the extraction of the metal. Till about twenty years ago aluminium was simply known as a chemical curiosity obtained in an impure state by Davy, and subsequently in a purer condition by Wöhler, Œrsted, and Bunsen, the last of whom prepared it by electrolysing the double chloride of aluminium and sodium or potassium whilst in a fused condition. Deville, about 1856, obtained the metal on a larger scale by acting on sodium with the vapour of aluminium chloride, prepared by heating strongly in a current of chlorine gas a mixture of pure alumina (oxide of aluminium) and carbon : although carbon alone will not remove oxygen from alumina, yet, when to the affinity of carbon for oxygen that of aluminium for chlorine is superadded, the oxide of aluminium is broken up, carbon oxide and aluminium chloride being formed. The process of Deville is now modified by employing the double chloride of aluminium and sodium, and fusing this compound with sodium and a flux of *cryolite* (a double fluoride of aluminium and sodium found in Greenland) : the sodium then takes away the chlorine from the aluminium, setting free the metal and forming sodium chloride. Cryolite alone and sodium may also be employed, the sodium in this case taking fluorine from the aluminium and forming sodium fluoride. The two essentials in the process are the preparation of sodium on a manufacturing scale, and the formation of a pure alumina free from silica, for the preparation of the aluminium and sodium chloride; this latter point is of the utmost importance, as, if not attended to, the aluminium finally produced is apt to contain silicon, which destroys the malleability of the metal and renders it crystalline and unsuitable for most of its applications. Clay, common enough everywhere, is therefore usually wholly unsuitable as a first raw material in the aluminium manufacture on account of its containing silica, generally in such a condition as to render it very difficult to separate from ordinary clay a pure alumina, excepting by the somewhat expensive process of manufacturing crystallized alum from the clay and precipitating alumina therefrom.

52. A variety of indurated clay, known as bauxite, is, however, found at Baux in the South of France and

elsewhere, from which it is not difficult to separate a pure oxide of aluminium by an inexpensive process, which was for some time worked by Messrs. Bell, Bros., at Washington near Newcastle :[1] this consists in heating in a furnace a mixture of ground bauxite[2] and soda ash (carbonate of sodium), whereby aluminate of soda soluble in water is formed, whilst the silica is either unacted on or is transformed into a silico-aluminate of soda not dissolved by water; on lixiviating the product with water the peroxide of iron present and the silico-aluminate of soda, &c., are left undissolved, and the aluminate of soda

[1] The production of aluminium at these works has been given up for several years, the manufacture not proving as remunerative as was at first anticipated. The chemical reactions taking place during the process are as follows :—
Production of sodium aluminate—
$$Al_2O_3 + Na_2CO_3 = CO_2 + 2AlNaO_2.$$
Decomposition of sodium aluminate by carbon dioxide (in solution)—
$$2AlNaO_2 + CO_2 + 3H_2O = Na_2CO_3 + 2Al(OH)_3.$$
Formation of alumina from the hydrate—
$$2Al(OH)_3 = 3H_2O + Al_2O_3.$$
Production of aluminium sodium chloride—
$$Al_2O_3 + 2NaCl + 3C + 3Cl_2 = 3CO + 2AlNaCl_4.$$
Manufacture of sodium from sodium carbonate—
$$Na_2CO_3 + 2C = 2Na + 3CO.$$
Extraction of aluminium from double sodium chloride—
$$AlNaCl_4 + 3Na = Al + 4NaCl.$$
Extraction of aluminium from cryolite—
$$AlNaF_3 + 3Na = Al + 4NaF.$$

[2] Average composition of bauxite :—

	Ure.	French.	Siemens. Austrian.	Irish.
Silica	2·8	3·5	6·3	3·5
Alumina	57·4	59·2	64·2	35·0
Ferric oxide	25·5	24·5	2·4	38·0
Calcium carbonate	0·4			—
Titanic oxide	3·1	12·8	27·1	2·0
Water	10·8			21·5
	100·0	100·0	100·0	100·0

CHIEF INDUSTRIAL APPLICATIONS.

is dissolved out ; by adding to the clarified solution the appropriate quantity of hydrochloric acid (or, better still, by impregnating it with carbonic acid gas) pure alumina is precipitated as a hydrate. This alumina is then dried, mixed with common salt (chloride of sodium) and charcoal, and formed into balls about the size of an orange ; vertical earthen retorts are then filled up with these balls, and heated to redness, whilst a stream of chlorine is passed through them ; the chloride of aluminium thus formed unites with the sodium chloride and distils over or sublimes as the double chloride, from which the metal is obtained by fusion with sodium.

The sodium required in this manufacture is simply prepared by heating to a very bright red heat a mixture of soda ash, charcoal or coke powder, and a little ground chalk; the action that ensues is of much the same character as that taking place in the distillation of zinc (§ 48) ; the sodium is reduced to the metallic state and distils over into peculiarly shaped condensers in which it collects, being preserved from oxidation by a layer of coal naphtha.

53. **Magnesium.**—The process employed for the production of this metal is in principle much the same as that used for aluminium, viz., decomposing the chloride (or double sodium or potassium chloride) with metallic sodium. To prepare the double sodium chloride, *magnesite* (native magnesium carbonate) or the artificial carbonate prepared from Epsom salts (sulphate of magnesia) by precipitation with an alkaline carbonate, or that extracted from *dolomite* (native double carbonate of lime and magnesia) by Pattinson's process,[1] is dissolved

---

[1] The dolomite is partially calcined, pulverized, and placed in a closed vessel with water; carbon dioxide gas is then forced in by

in hydrochloric acid, and an equivalent quantity of salt added; the whole is then evaporated to dryness and fused: or Epsom salts are decomposed by barium chloride and the solution of magnesium chloride separated from the precipitated barium sulphate and evaporated down after addition of salt. The double potassium chloride occurs naturally as the mineral *carnallite;* from either double chloride magnesium is readily reduced by simply fusing with sodium and a little fluor spar as a flux.[1] The metal thus obtained is purified by distillation in a current of hydrogen gas, the volatility of magnesium at a high temperature under these circumstances being sufficient for the purpose.

54. The chief use of the metal being to produce a brilliant light by its combustion, it is sent into the market mainly in the form of filings for mixture with pyrotechnic materials to add to their brilliancy when burnt, and as wire or ribbon, prepared by squirting (§ 74) the metal, heated till just fluid, through a conical orifice so that it solidifies as it is ejected, forming a continuous thin rod or wire; by passing this between powerful rollers it is flattened into a ribbon. In order to burn these wires or ribbons a spirit or other lamp is provided, to which is attached a small machine containing bobbins of the wire or ribbon connected with a clockwork arrangement so constructed that, on releasing a detent, the wires are slowly passed into the spirit flame in such a manner as to keep up a continuous magnesium

---

a pump; under a powerful pressure, magnesia is dissolved out as bicarbonate to the almost total exclusion of lime. The clear solution of bicarbonate of magnesia ("Dinneford's Fluid Magnesia") on heating gives magnesium carbonate as a precipitate and free carbon dioxide gas which is used over again for another operation.

[1] $MgNaCl_3 + 2Na = Mg + 3NaCl.$

flame. Another mode of attaining the same result is to allow the filings of the metal (mixed when necessary with sand, &c., to dilute them) to fall through a narrow tube into the spirit flame from a small hopper above; by means of a slide valve, the supply can be controlled or shut off altogether. An ingenious adaptation of this form of magnesium lamp has been proposed for the transmission of signals from ships, &c., by night. By opening the slide valve of a large arrangement on this principle for an instant only, a brilliant instantaneous flash of light is developed, visible in fine weather at long distances. By keeping the valve open for a somewhat longer period, a more continued blaze is produced; so that by properly regulating the intervals between the opening and shutting of the valve, any required signals may be telegraphed according to the Morse Code, the instantaneous flash representing the dot, and the somewhat more lengthened one the line or dash. The light given out by burning magnesium is very rich in actinic rays and is, consequently, of value for photographing in the absence of sunlight.

55. **Nickel.**—The principal minerals containing this metal are generally of so far complex a character that a succession of processes is requisite in order firstly to obtain a nickel compound free from other metals, and secondly, to reduce this compound to the metallic state. A considerable amount of nickel is extracted from "Speiss," a residuum left in the preparation of "smalt," which is an impure silicate of cobalt, prepared by fusing with siliceous matters and pearl-ash the roasted and washed cobalt and nickel arsenides and sulphides constituting the chief ores of these metals. During this process an arsenide of nickel (containing also sulphur, iron, copper, &c.) separates in the fused state and sinks to the

bottom of the smalt melting pot. From this speiss nickel is separated in the form of oxide by continued roasting, to drive off some of the arsenic and sulphur, solution of the residue in hydrochloric acid or aqua regia, careful precipitation of iron by means of a regulated addition of lime, with a little bleaching powder, separation from the filtered liquid of copper, bismuth, lead, &c., by passing through it a current of sulphuretted hydrogen gas, and finally precipitation from the clarified fluid of cobalt in the form of peroxide by addition of bleaching powder after neutralization by lime, and subsequently of nickel oxide by further addition of lime or soda to the filtrate from the cobalt precipitate. Other nickeliferous minerals admit of the nickel being separated by processes somewhat more simple than these, but generally consisting of the same kinds of operations, viz., solution in some appropriate acid, removal of foreign metals by sulphuretted hydrogen or other appropriate reagent, and, finally, conversion either into oxide or carbonate after separation from cobalt by means of bleaching powder or other chemical means.[1] In order to obtain the metal from these compounds they are mixed with flour or oil, &c., and strongly heated, when more or less compact lumps of reduced nickel are formed. If higher temperatures still be employed, the metal can be obtained in the fused state and cast into ingots; as thus prepared, the nickel contains a small quantity of carbon which communicates to it a higher degree of fusibility (as with iron, § 30). Occasionally the oxide is converted into oxalate; on heating this salt in a well-covered crucible metallic nickel is left. It has been proposed to smelt an alloy of copper and nickel by mixing together the

[1] *Vide* A. H. Allen, *Journal of the Society of Arts*, February 1878.

purified nickel oxide and copper oxide or carbonate and reducing the mixture by means of hydrogen, or carbon oxide, or by heating with charcoal powder or coaldust, the reduced alloy being ultimately fused and cast into ingots. Similar processes have been found to answer well in the case of other comparatively costly alloys; it is stated, however, that the nickel-copper alloy thus obtained does not answer so well for the manufacture of German silver (§ 103) as metallic nickel and copper prepared separately.

One chief application of nickel in the metallic state is as a coating for other more oxidizable metals, especially iron and steel, the nickel being deposited thereon by electroplating processes (§ 100).

56. It is noteworthy that nickel, cobalt, and iron, the three metals most magnetic in characters, and otherwise associated and allied in various respects, also possess the peculiar power of decomposing carbon oxide and setting free carbon therefrom at temperatures close to a low red heat.[1] This action is capable of going on to an indefinite extent, as the metallic oxide formed by the action becomes again reduced by a second portion of carbon oxide, thus—

(1) $xM + yCO = M_xO_y + yC$.
(2) $M_xO_y + zCO = M_xO_{y-z} + zCO_2$.
(3) $MxO_{y-z} + zCO = M_xO_y + zC$.

So that a continuous deposition of carbon and formation of carbon dioxide goes on in virtue of reactions (2) and (3). No other metals appear to possess this peculiar property.

57. **Antimony.** — Besides occurring as sulphide, occasionally intermixed or combined with the sulphides

[1] Lowthian Bell, *Chemical Phenomena of Iron Smelting*.

of other metals such as copper, iron, lead, silver, and nickel, antimony is found as a native alloy with silver and arsenic, and sometimes iron in addition, and also as oxide and oxysulphide; the chief ore, however, is the sesquisulphide Sb_2S_3, *stibnite*, which usually occurs in veins in calcareous or siliceous rocks, and sometimes imbedded in barytes (heavy spar or barium sulphate). From the gangue it is often separated by a process of liquation, the sulphide being readily fusible; the crude sulphide thus obtained is roasted to drive off arsenic and some of the sulphur; the mixture of oxide and sulphide thus obtained is then heated with charcoal and soda ash or potashes, whereby metallic antimony is formed which sinks to the bottom of the crucible, a fusible impure sulphide of antimony and sodium (or potassium) floating on the top: this is separated and utilized in the manufacture of antimonial compounds such as tartar emetic, Kermes mineral, &c. Occasionally the sulphide (separated from gangue by liquation or not according to the richness of the ore) is simply fused in a crucible with scrap-iron, whereby iron sulphide is formed and metallic antimony; or a combination of this method and the preceding is employed, the crude sulphide being roasted, and the impure oxysulphide thus formed being heated with potashes or soda ash, or, better still, salt-cake (sulphate of soda) and scrap-iron. If iron, copper, &c., be present in the metal thus obtained to an injurious extent, they may be removed either by powdering, mixing with antimony oxide and fusing so as to oxidize the impurities; or by treating the fused metal with small quantities of nitre, whereby the same result is brought about; another mode of purifying consists in fusing the metal in coarse powder with soda ash and a little antimony sulphide; in this way sodium sulphide is formed which sulphurizes and removes most of the impurities.

A method that has been found by Berthier to give good results as to yield and purity of metal is to heat in a suitable reverberatory furnace a mixture of antimony sulphide, iron oxide from the rolling-mills (smithy scales), soda-ash or salt-cake, and charcoal powder; or a mixture of antimony sulphide, scrap-iron, and salt-cake.

58. **Bismuth.**—For the most part this metal occurs in the native state disseminated throughout quartzose rocks; from these it is readily separated by simple liquation, the crude metal thus obtained being purified by fusion with a small quantity of nitre, whereby arsenic, iron, lead, sulphur, &c., are oxidized and removed. When other natural bismuth compounds are worked up, the most satisfactory process consists in obtaining the metal in solution, as nitrate, by treating the ores with nitric acid, addition of a large bulk of water so as to precipitate basic bismuth nitrate (subnitrate),[1] and reduction of this salt, after washing and drying, with charcoal in a crucible: a metal free from most of the ordinary impurities is thus obtained; it is, however, liable to contain small quantities of arsenic, the amount of which may be lessened or altogether removed by digesting the precipitated subnitrate with caustic soda solution, whereby the arsenic compounds carried down with the precipitate are mostly dissolved out: this method of treatment is also applicable for the refining of crude bismuth containing lead, iron, &c. Often silver is present in crude bismuth; to separate the two metals the mixture is dissolved in nitric acid and the silver precipitated as chloride from the solution before adding water to throw down the bismuth; or the whole is cupelled (bismuth answering

[1] Probably $Bi(NO_3)_3 + 2H_2O = 2HNO_3 + Bi(NO_3)(OH)_2$.

in the operation as well as lead), and the bismuth recovered from the oxide formed, the saturated cupels, &c., either by the wet process or by heating with coal, &c., to reduce the oxide. At Joachimsthal an ore containing bismuth, nickel, cobalt, and other metals is heated with scrap-iron and soda ash, with lime and fluor spar as fluxes: bismuth is thus reduced and sinks to the bottom along with a nickeliferous speiss cobalt(§ 55).

59. **Arsenic.**—The industrial applications of this metal in a reguline state are extremely few, the chief use of it being to form an alloy with lead for the purpose of shot manufacture, and a brilliant alloy known as *speculum metal* when combined with copper and tin in certain proportions: it is also employed for some few other purposes, such as the manufacture of opal glass. In the roasting of arsenical ores of various kinds, notably copper pyrites and nickel and cobalt ores, the fumes evolved in the furnace contain considerable amounts of arsenious oxide, As_2O_3; to collect this, and especially to prevent poisoning the air of the neighbourhood, these fumes are made to pass through a series of chambers opening one into the other, some arranged horizontally, and the last of the series vertically, constituting a kind of tower; in these the arsenious oxide is deposited along with other matters mechanically carried over, whilst the sulphur dioxide and other permanent gases escape by a flue from the highest chamber of the series. Fig. 19 represents the rotary calciner of Oxland and Hocking used for the roasting of arsenical pyrites for the purpose of preparing arsenious oxide therefrom: a cylinder of wrought iron a, a, a, a, such as an old boiler with the ends cut off, is lined with firebrick, ledges or shelves protruding from the lining radially towards the axis: the cylinder is supported on bearing wheels, b, b,

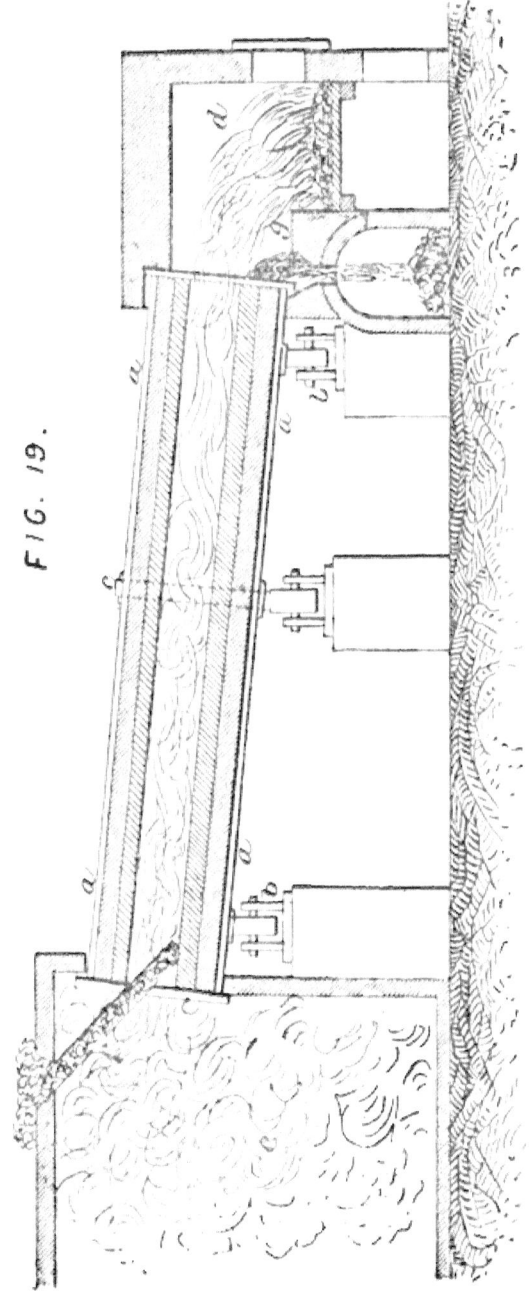

arranged on a gentle slope, and kept in motion (about fifteen revolutions to the hour) by means of a large cogwheel applied externally, c, and turned by water or steam power. The flame from a fire, d, plays into the cylinder at one end, and the gaseous products of combustion and of the roasting escape at the other into the first of the series of condensing chambers, e. As the cylinder revolves, the substance to be roasted is continually taken up by the shelves, and *poured* through the hot air inside in a shower when the shelves arrive near the top of the revolution, so as to keep up a continuous and very effective agitation of the whole mass. Fresh unroasted ore is continually added by a small shoot, f, at the upper end, the completely roasted ore escaping at the lower end at g. In this way the labour required in stirring, &c., in the ordinary calcining furnaces is saved, the operation proceeding automatically: when the ores operated on do not contain sufficient arsenic to render it worth while to apply condensing chambers, the cylinder can be adjusted simply to the flue leading to the chimney; in this way the calcination of tin ores (§ 47), and copper pyrites (§ 42), and zinc blende (§ 48), &c., can be effectually and economically carried out.

The arsenious oxide deposited in the condensing chambers is purified by a process of sublimation before sending into the market; the majority of arsenical compounds in use are made from the purified oxide direct without passing through the reguline condition at all. From the oxide, the metal is readily obtained by mixing with charcoal powder, or with black flux (an intimate mixture of potassium carbonate and charcoal prepared by calcining tartar in a closed vessel) and heating, when the metal sublimes. It can also be obtained in a less pure condition by heating arsenical pyrites sulphide of arsenic and iron) mixed with scrap-iron in a retort; sulphide of iron is thus formed, whilst the arsenic is volatilized and condenses in a receiver placed so that the vapours may sublime into it.

60. **Manganese**.—Practically, manganese is only prepared in the form of alloys, of which the most important, *spiegeleisen* and *ferromanganese*, are obtained by smelting a manganiferous iron ore, or a mixture of manganese oxide and an iron ore, in blast furnaces, &c., precisely as pig-iron is prepared (§ 32). In this way a highly carboniferous pig-iron is obtained, containing more or less manganese according to the ore employed. When only a few per cents. of manganese are present, the substance is usually termed *spiegeleisen*, on account of the fractured surfaces of the pigs exhibiting brilliant mirror-like planes. When the percentage of manganese is upwards of 7 or 8 per cent. (occasionally running up to 40 or 50 or more per cent.), the term "ferromanganese" is usually assigned to the alloy. No distinct line of demarcation between spiegeleisen and ferromanganese can, however, be laid down, the one insensibly grading into the other.

Other manganese alloys have been proposed for special purposes, but have not as yet come into anything like general use; recently a manganese substitute for German silver has been prepared, consisting of about 8 parts manganese, 4 zinc, 34 copper, and 1 iron; whilst a manganese-bronze has also been brought out as a cannon metal, and is stated to possess great toughness, and to be equal to the best wrought iron in firmness and extensibility, whilst it can be easily forged, rolled, and drawn at red heat. If these statements are substantiated on further trial, no doubt the new alloy will come into extensive use.

61. Of the other oxidizable metals none have any special importance as regards their industrial applications in the reguline state, and very few of them are ever prepared in that condition save by the manufacturers of the rarer chemicals. *Cadmium*, occasionally employed for the production of very fusible alloys (§ 88), is usually obtained along with zinc, and is extracted as indicated in (§ 50). *Tungsten* has been recently proposed by Dr. Versmann, as a substitute for nickel in German silver, a highly sonorous alloy being obtained by reducing simultaneously tungstic and copper oxides, and fusing the resulting alloy with zinc or brass ; the same metal has been suggested as a hardening and ameliorating ingredient in steel, as have also *chromium* and *titanium* (§ 41). Few indeed of the remaining metals of this class are of any large industrial importance even in the state of compounds, *potassium*, *sodium*, and *calcium* being the main exceptions. Certain of the compounds of *barium*, *strontium*, *chromium*, and *cobalt* are also of considerable importance for a few special purposes and manufactures ; *vanadium* compounds have been recently introduced for developing certain colours in calico-printing, &c., and *cerium* salts have been used medicinally, whilst *molybdic* oxide is a valuable analytical reagent; but for the most part the remainder of the oxidizable metals and their compounds have as yet received no industrial applications.

CHAPTER V.

PHYSICAL PROPERTIES OF METALS.

62. **Lustre.**—One of the most characteristic properties of metals is the power possessed by them when in more or less compact masses of acquiring (by polishing, pressure, or other mechanical treatment) such a condition of surface that light incident thereon is for the most part again reflected, whereby a peculiar glistening appearance is presented, known as the *metallic lustre*. When fused, lumps are cut with a sharp instrument, or when the metals form crystalline masses which are broken across, the severed surfaces exhibit this peculiar feature; and in many cases when the metal is obtained by chemical action from certain of its compounds, so as to form a deposit on a glass surface (*e.g.* silver), or if the vapour of the metal be condensed on such a surface (*e.g.* arsenic), the glass presents the well-known appearance of a *mirror* of more or less brilliancy according to the nature of the metal, and the way in which it is precipitated on the surface. When prepared by certain chemical processes, the metals often present the appearance of lustreless minute particles, generally black when very small. Metals in this "spongy" condition, however, can often be made to exhibit a considerable amount of brilliancy by

strongly compressing a portion of the dry powder against a polished steel or agate surface, when the powder more or less agglutinates into cakes exhibiting lustre on the sides that were next to the polished surfaces ; or by simply rubbing in a smooth mortar with a pestle a little of the powder, brilliant streaks are produced by the pressure. Gold, platinum, and silver in particular exhibit this property, the ordinary metals having a strong tendency, when in a fine state of division, to oxidize or rust on the surface ; and this greatly interferes with the coherence of the compressed masses, and with their brilliancy when rubbed or polished. If a little spongy silver (prepared by boiling pure silver chloride with sugar and caustic soda, thoroughly washing, and drying) be placed between a pair of dies and compressed by a coining press, a slightly brittle but still coherent coin will be obtained, exhibiting considerable brilliancy on the surfaces that were in contact with the polished portions of the die-surfaces ; and an analogous result is obtained if precipitated gold of a dull brown shade (thrown down by ferrous sulphate from gold chloride solution) be employed instead of silver.

The peculiar lustre characteristic of metals is, however, not wholly confined to these substances ; certain minerals, *e.g.* galena, and some of the non-metals, *e.g.* graphite, occur in nature in masses exhibiting a metal-like lustre when fractured, whilst selenion, silicon, and other non-metals can be artificially prepared in states where they exhibit a closely similar lustre, as can also various compound substances, *e.g.* mosaic gold (tin disulphide).

The effect of pressure in heightening the brilliancy of a metallic surface is well seen in the industrial process of *burnishing;* when a layer of metal, such as gold or silver, is deposited by electrical or chemical means on the surface of an object to be silvered or gilt, it usually happens that the freshly covered surface is more or less deficient in lustre, or, as it is technically termed, "dead"; by the skilled application of pressure with a burnisher of polished steel or stone the dead surface is compressed and made to shine with brilliancy. For the final touches on electrosilvered or gilt articles, and for the gilding on china, &c., a burnisher is preferred made of "bloodstone," a compact variety of hæmatite. The ordinary household processes of cleaning and polishing steel and other metallic articles partly depend on the pressure exerted on the surface, although the main action consists in abrading, by means of a fine crystalline powder (usually prepared chalk, or peroxide of iron prepared by calcination of green vitriol or by levigation of burnt pyrites, &c.), the particles of rusted or tarnished metal, so as to display the underlying pure metallic surface.

63. Owing to the influence of the air, moisture, vapours arising from putrefaction, &c., metallic surfaces, even when highly polished and brilliant, become more or less rapidly tarnished, so that the power of reflecting light is to a considerable extent lost. Before the invention of glass, polished metallic surfaces were employed as *mirrors;* and for reflecting telescopes such surfaces are still in use. Now, however, it is usual to employ as mirrors glass surfaces, behind which a thin coating of some lustrous metallic mass is placed, so that the smooth surface of the glass at once determines the peculiar reflective power of the metal applied to it, and preserves the metal from mechanical injury and from the corrosion of the air. For this reason these household appliances are ordinarily termed "looking glasses," although strictly speaking it is not the glass that is the essential part.

Three principal methods of applying these metallic substances to glass are in use; the best plate-glass mirrors (perfectly plane surfaces) are prepared by spreading out on a table surrounded with a deep groove or gutter, and capable of being raised on hinges so as to be placed at any angle with the horizon, a sheet of tinfoil, and smoothing it with a soft brush; mercury is then poured on and gently rubbed over the tinfoil with a hare's-foot or a roll of flannel so as to penetrate and brighten the tin; more mercury is then poured on, and the surface cleansed from dross, &c.; finally, the perfectly clean sheet of glass is dexterously slid over the brilliant mercurial surface in such a way as to avoid enclosing any particles of dust or air-bubbles between the metal and glass. The table is then slightly raised at one end, so that the surplus mercury may gradually run off and be caught in the gutter; and the slope is increased daily, a piece of flannel being placed on the glass with weights on it to facilitate the draining off of the mercury. After two to four weeks, according to the size of the plate, the mirror is complete, the tin amalgam having then completely set, and being tolerably firmly adherent to the glass, although easily rubbed off and scratched on account of its slight tenacity. To preserve the back of the mirror from injury a suitable wooden frame is provided, in which the whole is fixed, when a finished mirror is the result.

64. For curved surfaces, such as the insides of globes, flasks, &c., for ornamental purposes, a somewhat different plan is employed: a fluid or semi-fluid amalgam capable of adhering to glass is poured into the vessel to be "silvered," and shaken about therein until the inner surface is covered with a film of the composition; the surplus amalgam is then poured out and used for other similar objects. A mixture of one part each of lead, tin, and bismuth, with two parts of mercury, answers well, the

CHIEF INDUSTRIAL APPLICATIONS.

mixture being made perfectly fluid by slightly warming it before pouring into the vessel to be silvered. It is noticeable that an alloy of three parts of potassium and one of sodium is fluid at the ordinary temperature, being the only metal or mixture of metals known possessing this character, mercury, solutions of metals in mercury, and the newly discovered metal gallium (under certain conditions) excepted. This mixture possesses the power of adhering to the inside of a glass bottle forming a well-defined mirror; fused stereotype metal, plumber's solder, and analogous alloys (§ 103) can also, with careful management, be poured into hot glass bottles and shaken over the surface so as to form mirrors of considerable brilliancy, although this method is never practically used.

A method which has of late years come largely into use for silvering mirrors of various kinds, and notably the reflectors of telescopes and lighthouses, is based on the power of certain chemical reagents to throw down silver in the metallic state from certain of its solutions, &c., the reduced silver in many cases adhering firmly to the surface of the vessel in which the action takes place, or to objects immersed in the liquid. Thus, if calcium tartrate in a moist state be placed in a glass vessel with a crystal of silver nitrate and a drop of ammonia solution, and the mixture cautiously heated, and made to flow successively over the whole inner surface of the glass, a fine mirror may be developed. Aldehyde, oil of cloves and other essential oils, grape-sugar, and some other organic substances, may also be employed as reducing agents, especially the first substance.

65. If a "mirror" (*i.e.* a glass surface with a brilliant metallic film behind) be carefully examined, it will be found that in most positions it will give a double image

of any object reflected, one image being usually more brilliant than the other. Fig. 20 illustrates how this is brought about; a ray of light from an object at a, strikes the glass surface at b, and is reflected to the eye of the observer at P, so that an image is seen situated at m. Another ray of light incident on the glass at a point c, is partly reflected along cf, this portion of the ray consequently never reaching the eye at P at all; the rest

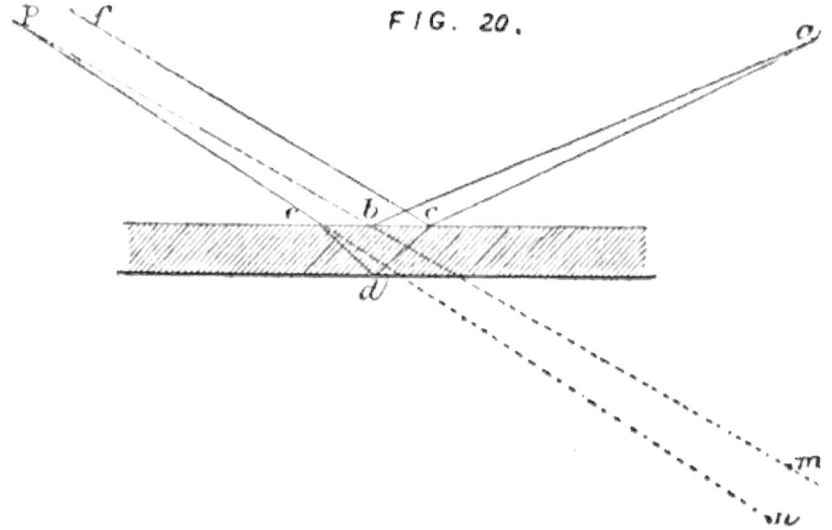

FIG. 20.

of the ray enters the glass, being refracted along cd; at the junction of the glass and metallic surfaces reflection takes place along de, and at e the ray is refracted along eP, thus also reaching the eye of the observer, but necessarily causing the image formed to be seen apparently situated at n, a point different from m. The relative quantities of light passing along eP and bP (that is, the relative brightnesses of the two images) depend on the degree of obliquity of the incident light c; the greater the angle $a\,b\,P$ (*i.e.* the more obliquely the light falls on the mirror), the brighter is the image at n. The power of glass thus to reflect light to a considerable extent without any metallic film behind is utilized in the illusion known popularly as "Pepper's ghost," which consists simply of a large pane of glass sloping forwards from the stage at an angle of about 45° (Fig. 21.) Objects such as $A\,B$, placed between the footlights E, and the pane of glass F in a horizontal position, and strongly illuminated, will produce to a spectator in front at P, a

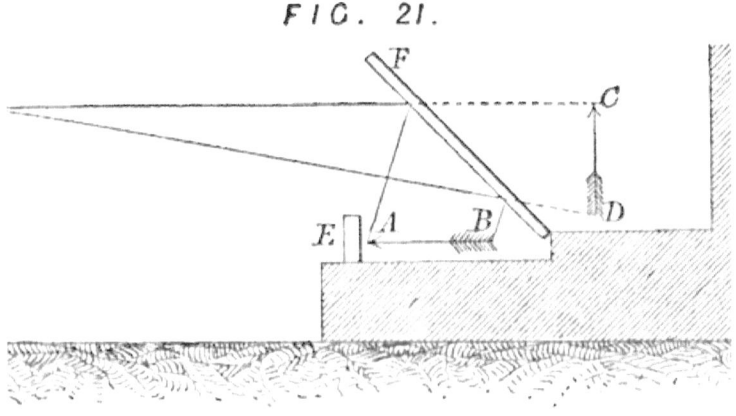

FIG. 21.

virtual image or "ghost," apparently situated at $C\,D$, the illusion being heightened by hiding, by means of screens, all the apparatus in front of the pane from the audience, and darkening that part of the stage behind the pane, the real objects furnishing the ghosts being placed on a dead-black ground. When the lights E are extinguished, and other lights illuminating the stage behind the pane turned on, the ghosts disappear, whilst the real actors at $D\,C$ on the stage behind the pane become visible *through* the transparent glass.

66. Colour by Reflection and Transmission.

—As a rule, metals reflect visible light of all degrees of refrangibility nearly alike, *i.e.* most metals appear of a *white* colour. Copper, however, possesses the power of reflecting red rays more powerfully than others, and consequently appears red : similarly, gold possesses a bright yellow or orange colour, and the alkaline-earthy metals calcium, barium, and strontium, appear slightly yellow. Brass, aluminium bronze, and other alloys of copper possess a rich yellow colour, the shade depending on the composition; as a rule, the alloys of a coloured and a colourless metal exhibit a regular gradation of tint, the colour becoming less intense as the percentage of colourless metal increases. Although the light reflected from polished surfaces of most metals is nearly white, yet frequently there is a slight tinge of some colour ; thus whilst tin, silver, platinum, and others have a nearly pure white colour, and hence appear alike when equally burnished, lead and zinc have a bluish shade, and iron and arsenic a greyish hue ; these faint tints are best seen by repeatedly reflecting light from the metallic surface, as when a tube polished internally, and open at both ends, is looked into obliquely. The deepening of the yellow tint exhibited when light is once reflected from a gold surface to a red-orange by repeated reflections is readily seen by looking obliquely into an empty metal tankard gilded internally. The pigments and colouring materials largely used in the arts under the name of *bronze powders*, are mainly divers coloured alloys stamped and ground to fine powders, and in some cases subjected subsequently to heating and sulphurising processes so as to develop peculiar shades ; *tinsel* is usually paper overlaid with a thin film of silver or other white alloy, and coated over with a transparent coloured varnish, or is paper coated with a white or bronze powdered metal, and rolled till the surface is brightly lustrous.

67. Few metals can be reduced to so fine a degree of tenuity as to allow light to pass through them readily; when this can be done it is usually found that one kind of light is absorbed more readily than others, so that the transmitted light is coloured, being deficient in the more absorbed rays. The transmitted light is sometimes complementary to the reflected colour, or nearly so, but not invariably: thus gold can be reduced by beating to leaves of thickness not exceeding $\frac{1}{260,000}$ inch (§ 72), and in this condition permits a green light to pass through, the colour being dependent on the amount of silver added to the gold, the light being more inclined to violet with much silver: by passing electric sparks in vacuo through glass tubes into which metal wires are fused, the glass becomes coated with a continuous film of metallic particles detached from the wires by the sparks; a film of gold thus prepared transmits a fine green light, whilst silver gives a beautiful blue shade. Copper transmits a dull green, and platinum a bluish-grey: zinc and cadmium furnish a deep bluish-grey, whilst iron films transmit a tint nearly neutral, but slightly brownish.

68. **Density.**—Most of the metals used in the arts in the free state are of considerable density, aluminium being by far the lightest, a circumstance which, together with its considerable strength and power of resisting the tarnishing effects of the air, renders it peculiarly suitable for numerous purposes: the draw-tubes of telescopes, opera glasses, &c., and the graduated circles of surveying instruments, &c., are often made of this metal for these reasons. According to the way in which a piece of metal has been obtained, its density will vary somewhat, being increased by hammering or any mechanical action

which forces the particles together, *e.g.* wire drawing or sheet-rolling. The following table gives the numerical values of the average densities of most of the more im-

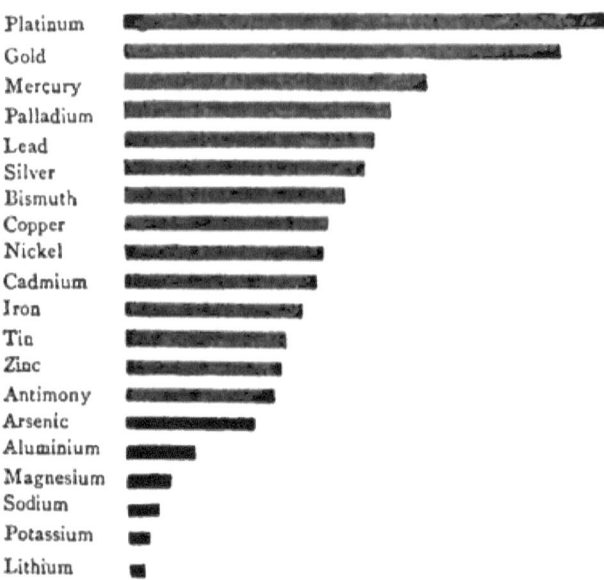

FIG. 22.

portant metals, whilst Fig. 22 exhibits the same numbers graphically, the lengths of the lines applied to each metal being in the ratios of these numbers.

Specific Gravity of Metals (Water = 1).

Platinum	21·5	Iron		7·8
Gold	19·3	Tin		7·3
Mercury	13·6	Zinc		7·1
Palladium	11·8	Antimony		6.7
Lead	11·3	Arsenic		5·6
Silver	10·6	Aluminium		2·6
Bismuth	9·8	Magnesium		1·8
Copper	8·9	Sodium		0·97
Nickel	8·8	Potassium		0·86
Cadmium	8·7	Lithium		0·59

These numbers necessarily represent the numbers of grammes weighed by one cubic centimetre of each metal; when multiplied by 1,000 they represent approximately the number of ounces per cubic foot. Independently of the changes in density produced by differences in the state of physical aggregation of metals, alterations of temperature of course affect this property, since expansion is caused by increase of temperature, and consequently a given bulk of metal will weigh less at a high temperature than at a lower one.

69. It is noticeable that the specific gravity of an alloy always approximates more or less closely to that calcutated on the assumption that the constituent metals hang together side by side without interference: *i.e. the bulk of an alloy approximates to the united bulks of the constituents.*

Thus an alloy of equal *volumes* of platinum and aluminium would, on this assumption, have a specific gravity of 12·05, the calculation running thus:—

		Grammes.
100 cubic centimetres of platinum . . . . weigh		2,150
100 ,, ,, ,, aluminium . . . ,,		260
200 ,, ,, ,, the alloy . . . . ,,		2,410

or, 1 cubic centimetre of the alloy weighs $\frac{2410}{200}$ grammes $= 12\cdot05$

On the other hand an alloy of equal *weights* of these two metals would have a specific gravity of 4·64.

100 grammes of platinum occupy $\frac{100}{21\cdot5} = 4\cdot65$ cubic cents.

100 ,, aluminium ,, $\frac{100}{2\cdot6} = 38\cdot46$,, ,,

200 ,, the alloy ,, $43\cdot11$,, ,,

whence 1 cubic cent. weighs $\frac{200}{43\cdot11} = 4\cdot64$ grammes.

In the case of most alloys this proposition is not exactly true, there being generally more or less expansion or contraction during the mixture of the constituents, so that the specific gravity of the alloy becomes either raised above or lowered below that thus calculated; sometimes this difference is somewhat marked, but usually it is not great enough materially to vitiate the generality of the above rule.

70. **Crystallisability.**—Some few metals readily assume the crystalline condition on slow cooling after fusion; notably this is the case with bismuth; others do not readily become crystalline, and on this property depends much of their usefulness, as a crystalline texture denotes comparative brittleness. Wrought iron occasionally passes from its normal *fibrous* texture to a crystalline state; this effect appears to be brought about by long continued vibration and subjection to blows, and has been accordingly the cause of various accidents, *e.g.* when a railway axle having become crystalline through long use, breaks owing to the diminished strength thereby caused. Many metals can be obtained crystalline when deposited from a solution by galvanic action; thus lead and silver when slowly precipitated by other metals occur in finely developed crystalline arborescences, the formation of which constitutes the familiar experiments of the "lead-tree" and "silver-tree" or *Arbor Dianæ*. Many "native" metals (*e.g.* copper) occur crystallised, being presumably produced by actions of this kind. Frequently the presence of a minute quantity of a foreign metal greatly promotes the crystallisation of a metal which becomes crystalline only with difficulty when perfectly pure.

71. **Malleability and Brittleness.**—Many metals are sufficiently devoid of the character known as *brittleness*, or tendency to fly to pieces when struck or

pressed, to be hammered out into thin leaves, or rolled out into thin sheets; some metals are readily "malleable" at one temperature but brittle at another: in all cases the presence of minute amounts of other metals or non-metallic impurities exerts a marked influence on the degree in which this quality is possessed. Zinc is crystalline and brittle at ordinary temperatures, but can readily be rolled into thin sheets at 100°—150°, whilst at about 200° and upwards it again becomes brittle, and can be powdered in a mortar; sheet zinc for gutters, &c., is therefore rolled hot. Gold is rendered brittle by the presence of traces of antimony; and similar effects on the malleability of many metals are produced by small admixtures with certain other metals, &c.

The ten chief metals (§ 4) all possess the power of being extended into thin sheets under the rolling press, and into leaves of greater or less tenuity under the hammer, but to very different extents; lead and zinc falling short of the others, the former on account of its extreme softness and consequent want of tenacity when reduced to thin leaves, the latter on account of the difficulty of working very thin sheets of the metal at the temperature at which its malleability is most marked. They may be thus arranged in order of malleability, the first two being nearly equal:—

GOLD, SILVER, COPPER, PLATINUM, IRON, ALUMINIUM, TIN, ZINC, LEAD, MERCURY (solid).

It is very probable that mercury (in a frozen state) should be ranged higher up in the series; palladium, like platinum, is highly malleable, though less so than gold or silver, which possess the property to an extraordinary extent; in the ordinary process of gold-beating leaves are obtained so thin that one grain of finished leaves covers at least thirty-five square inches, whilst the extension of this metal can be pushed much further, so that coherent leaves of $\frac{1}{370,000}$ inch in thickness have been obtained, one grain of gold covering about seventy-five square inches: the usual thickness of English gold leaf is about double this, or $\frac{1}{200,000}$ inch. Silver can similarly be beaten until a given weight of metal is even more extended, one grain covering ninety-eight square inches. The leaves however are not quite so thin as gold leaves, on account of the lower specific gravity of silver. Iron has been beaten into leaves of $\frac{1}{2500}$ inch in thickness.

72. The manufacture of thin leaf gold (*gold-beating*) is carried out in the following way. Gold is alloyed with small quantities of other metals according to the colour required in the finished leaf; thus there were exhibited in the 1851 Exhibition by Messrs. Marshall leaves of twelve colours grading from red to nearly white, and designated as—red: pale red: extra deep: deep: orange: lemon: deep pale: pale: pale-pale: deep party: party: and fine gold. The deeper colours are obtained with gold alloyed with from twelve to sixteen grains of copper per ounce and no silver; the middle ones with six to eight grains of copper and twelve to twenty of silver; and the paler ones with from two to twenty grains of silver, copper being omitted. As a curious fact, if silver be added to alloy containing more than eight or ten grains of copper the malleability is sensibly diminished, although no marked ill result is brought about by the addition of silver to fine gold or to gold containing only small quantities of copper. Ordinary gold leaf, such as is used for decorative purposes, contains about twelve grains of silver and nine of copper per ounce. The alloy is heated in crucibles to some-

what above its melting-point, cast into ingots, and rolled into ribbons about one-and-a-half inch wide and of such thinness that an ounce extends to a length of about ten feet: no marked difference in malleability is noticeable whether the cast ingot be cooled quickly or slowly. The ribbon is then cut into pieces weighing about six grains each, which are piled between sheets of vellum or parchment paper to the number of about 160 or 170: the *cutch* thus produced is beaten with a seventeen-pound hammer for about twenty minutes, at the end of which time the metal has become extended nearly to the size of the parchments (some three inches square). The rough squares thus formed are quartered and piled again between sheets of prepared gut (*goldbeater's skin*) forming a *shoder* (or *sholder*) consisting of some 700 pieces of metal, the contents of a cutch after quartering being all placed in the same shoder. The beating is then carried on for about two hours with a nine-pound hammer: when the leaves have become extended to the dimensions of the shoder (some four-and-a-half inches square) they are again quartered by a tool made of bamboo sharpened to a cutting edge. These quarters are again piled between fine skins five inches square forming a *mould*, the contents of one shoder filling three moulds (about 900 leaves to the mould). Finally, the mould is beaten for about two hours more till the metal extends to the edges of the skins and here and there flows over: from the finished leaf thus produced squares of three inches and three-eighths are cut from the central portions by a bamboo tool, and piled in a " book " made of soft paper rubbed over with red ochre or red chalk. During the earlier part of the beating the blows are directed mainly towards the centre, which causes cracks and rents towards the edges of the leaves; these cracks, however, become perfectly closed up again subsequently as the blows fall on the other portions, the edges of the cracks *welding* together perfectly, so that the finished leaf exhibits no trace of them. The leaves first begin to show light through when the thickness is about $\frac{1}{150000}$ inch, the colour being green with gold containing little or no silver, but verging towards pale violet when much of the latter metal is present. The beating is rarely continued until the leaves are less than $\frac{1}{200000}$ inch in thickness, as the saving of precious metal hardly compensates for the extra labour and the greater waste (through spoiled leaves, and the "pouring over," at the edges of the mould): moreover the thinnest possible leaf does not "cover" so well, being more translucent. Fine gold beats as well as, but not better than that containing small quantities of alloy, whilst it is superior in welding power, so that the leaves are more apt to stick together when one part touches another, and so on. The "mould" skins when old are generally employed for the "shoder," in which the excellence of the skin surface is not of so much moment.

73. **Ductility.**—Although the property of being drawn into wire is closely allied to that of being rolled or hammered into foil and leaves, yet the two are not necessarily possessed to equal extents by the same metal; gold, silver, and platinum are pre-eminently "ductile," whilst copper and iron are but little inferior to them in this respect. Aluminium and zinc can be obtained in tolerably thin wire, whilst lead and tin have so little cohesion that they cannot be drawn beyond a very limited degree of fineness. On the small scale, wires are readily obtained by casting the metals into thin pencils,[1] slightly pointing

[1] For metals of moderately-low melting-points the fused substance may be drawn up into a hot thin glass tube or pipe-stem by suction, and allowed to solidify therein. By fusing the metal in the bowl of a tobacco-pipe and tilting this so that the stem is inclined down-

the ends of these and passing them into a funnel-shaped hole in a steel plate (*draw-plate*) of suitable size, gripping with pliers the protruding pointed part, and forcibly pulling the whole bar through the hole, the process being then repeated with a slightly smaller hole. Fig. 23

FIG 23

represents the section of the draw-plate through the holes; a series of holes are generally worked in the same plate gradually diminishing in diameter from one end of the plate to the other, the complete perforated plate being often termed a "jigger." To obtain greater steadiness in the pull, the pliers should be attached to a band or cord which is gradually wound up on an axle by a handle, the pliers being so constructed that the greater the force required to draw the wire through, the more firmly they grip the end of it: this is easily effected by turning up the handle ends (the plane in which the jaws open being horizontal) and passing over them a triangle of iron to the base of which the band is attached (Fig. 24): the greater the strain on the band, the more firmly is the wire held: the draw-plate is kept in position by being pressed against two vertical projections or "chucks." It is generally necessary to "anneal" the wire from time to time, otherwise it becomes hard and more or less liable to crack or break after having passed through a certain number of holes: "hard-drawn" (or unannealed) wires, however, are usually considerably

---

wards, the molten metal can often be made to form a rough wire or thin rod in the stem, readily obtainable by breaking away the pipe-clay after cooling.

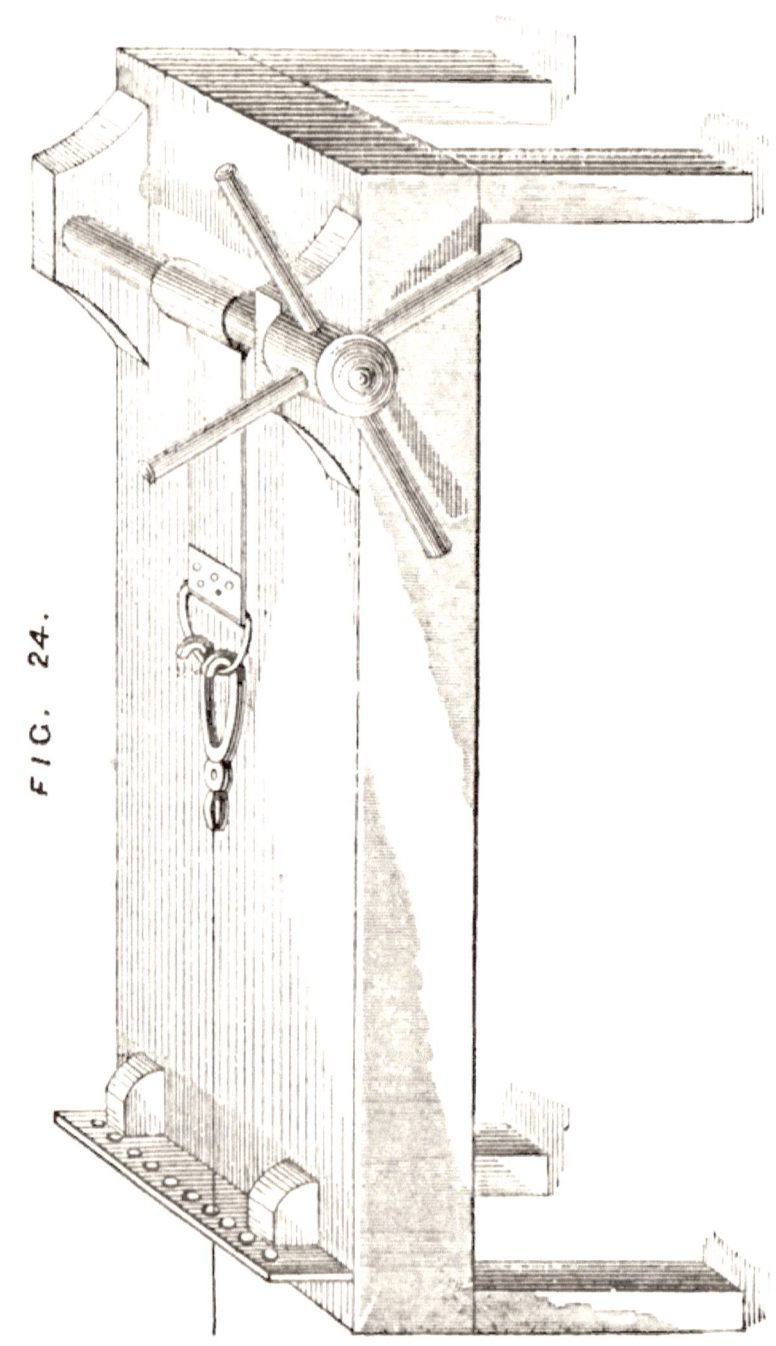

FIG. 24.

more capable of resisting tensile strains than the same wires after annealing (§ 76).

In drawing wire on a manufacturing scale, the process is just the same in principle, only, instead of drawing the wire through the draw-plate by hand by means of a wheel and axle, &c., the wire is pulled through by hand with pliers for a foot or two, and this portion then fastened to a revolving drum which then pulls the rest of the wire through, coiling up the drawn-out portion on the drum; the wire is then passed through the next smaller hole, being uncoiled from the first drum, and coiled again on a second in so doing, and so on until drawn to the required degree of fineness. In this way great lengths of wire are drawn at one operation.

74. Wollaston succeeded in obtaining wires of platinum, gold, and iron of excessive tenuity by first drawing the metals into fine wire, and then casting round this wire a cylinder of another metal, and drawing the compound cylinder again to the utmost possible extent; both the outside metal and the internal wire were thus elongated together; finally, the outside metal was dissolved off by some appropriate solvent, and thus the internal wire was left. To prepare the platinum and gold wires silver was used for the external cylinder, the silver being finally dissolved off by nitric acid. To make the iron wire, silver was also used for the cylinder, but was ultimately dissolved off by means of mercury. In this way a platinum wire was obtained less than $\frac{1}{30000}$ inch in diameter.

Some metals require to be heated in order to acquire sufficient softness to enable them to be readily drawn into wire; thus with aluminium. Others are in practice often used in a heated state to facilitate the operation, although the heating is not essential to the ductility of the body. Thus with steel and iron wires or rods of considerable thickness, the "drawing" operation being modified into a kind of rolling, the hot metal being passed between grooved rollers of successively smaller and smaller grooves.

Those alloys which are brittle are necessarily non-ductile, but alloys which are malleable are usually more or less ductile. Brass, a particularly malleable alloy, is also excessively ductile, wires of this alloy having been drawn so fine by means of an ordinary drawbench that seven feet of wire only weighed one grain. It is noticeable that "virgin" brass (made by mixing fresh copper and zinc together, not by remelting old alloy, &c.) is much more ductile than other kinds of brass, even though apparently of the same chemical composition.

Many metals which are only drawn into wire with difficulty, can be obtained in a wire-like form by an ingenious process called "squirting." The metal is melted and allowed to cool, and when close upon the solidifying point is forced by mechanical means (hydraulic pressure, &c.) through an orifice of the requisite diameter; the metal, coming in contact with the air, is quickly chilled, and solidifies to a wire or rod, which is wound on a drum as fast as it is prepared. By an appropriately formed jet or orifice a long continuous *tube* may be squirted instead of a rod. In this way the ordinary tin, lead, and "compo" tubing used for gas, water, spirits, &c., is manufactured, as is also magnesium wire. Rifle bullets are made by squirting lead into rods half an inch thick or so, from which portions are cut off and squeezed into shape by machinery.

75. **Tenacity.**—The more crystalline a metal or alloy is the less is its power of resisting strains and stresses of various kinds; the more fibrous in structure the better will it resist strain, especially in the direction of the fibres.

By forming metals into wires of equal dimensions, and then determining the weight requisite to break these wires, the differences in tenacity exhibited by metals and alloys may be readily demonstrated. A convenient apparatus for this purpose is made of an iron tripod six or seven feet high (Fig. 25), the legs of which are stayed together at the bottom and in the middle; from the top of the tripod is suspended by a stout hook a dynamometer or spring balance furnished with a hook at the bottom, whilst about half way up the tripod is affixed a horizontal axle, supported by the stays in such a position that the centre of the axle is perpendicularly beneath the hook of the dynamometer. This axle is provided with a winch, and round it is coiled a stout rope or leather band with a hook at the end. The wire to be tested is formed into a ring about three or four inches in diameter, the ends being intertwisted and soldered together; the hooks attached to the bottom of the dynamometer and to the rope are then inserted in this ring, and the handle turned so as to wind up the rope and stretch the ring until its form becomes a narrow oblong. The tension is then increased by winding the rope until the wire breaks; the reading of the dynamometer is noted by an assistant at the moment of rupture.

In this kind of way the order of tenacity of the metals is found to be as follows :—

    25 Iron.
    16 Copper.
    14 Platinum.
    12 Aluminium.
    10 Silver.
     8 Gold.
     7 Zinc.
     1·5 Tin.
     1 Lead.

FIG. 25.

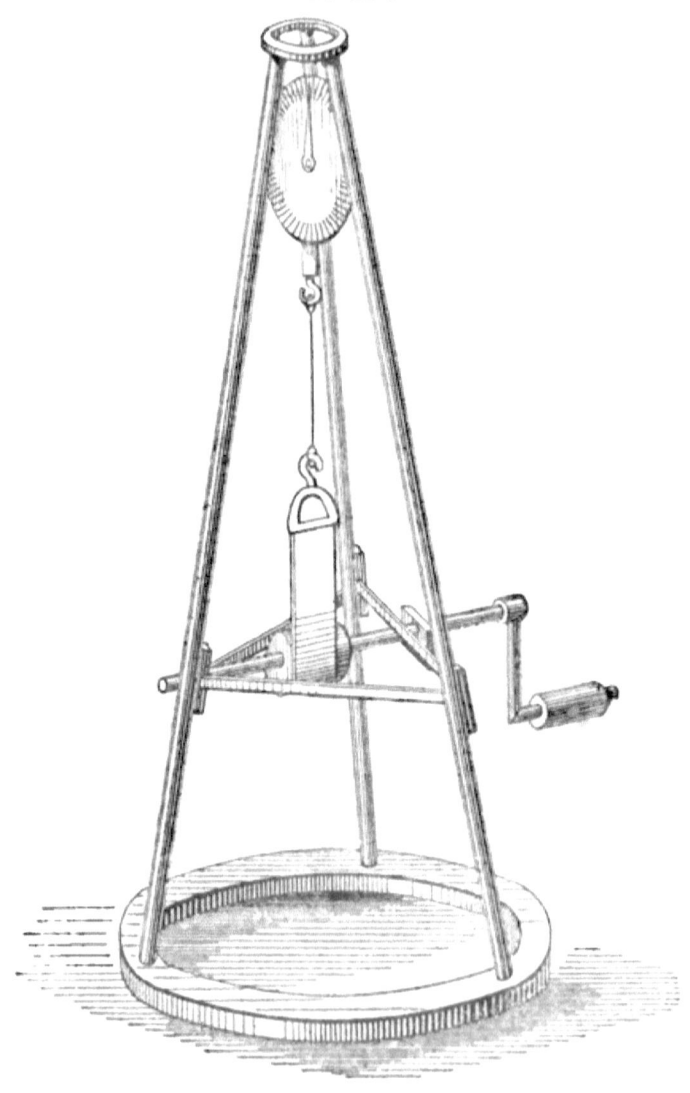

76. These numbers represent the relative average breaking strains of wires of the same dimensions. The actual value obtained with any particular wire, however, is largely influenced by the purity of the metal, by the character of the strain (whether applied suddenly, or slowly and gradually), and by the physical condition of the wire (whether hard drawn or annealed). Thus, Wertheim[1] obtained the following values representing the weights in kilogrammes required to break wires one square millimetre in section :—

	Gradual strain.	Strain suddenly applied.
Cast steel, hard drawn	—	83·8
,, annealed	65·7	—
Piano wire (steel)	70·0	99·1
,, annealed	40·0	53·9
Iron wire, hard drawn	61·1	65·1
,, annealed	46·9	50·3
Copper wire, hard drawn	40·3	41·0
,, annealed	30·5	31·7
Platinum wire, hard drawn	34·1	35·0
,, annealed	23·5	27·7
Palladium wire, hard drawn	—	27·2
,, annealed	27·4	—
Silver wire, hard drawn	29·0	29·6
,, annealed	16·0	16·5
Commercial zinc, drawn	12·8	15·8
,, annealed	—	14·4
Pure zinc, cast	4·5	—
Gold wire, hard drawn	27·0	28·4
,, annealed	10·1	11·1
Cadmium, drawn	2·24	—
,, annealed	—	4·8
Lead, cast	1·25	2·21
,, drawn	2·07	2·36
,, annealed	1·80	2·04
Tin wire, hard drawn	2·45	3·00
,, annealed	1·70	3·62

[1] *Annales de Chimie*, [iii.] xxii. 440.

Somewhat different values have been obtained by other experimenters, the differences being probably mainly due to the presence of minute traces of impurity in the metals examined, &c. Thus Matthiessen gives the following values, somewhat different from the above. The numbers represent the number of pounds weight required to break hard-drawn double wires, No. 23 gauge.

Steel	above 200
Iron	80–90
Platinum	45–50
Silver	45–50
Copper	25–30
Gold	20–25
Tin	under 7
Lead	,, 7

77. Wires sometimes vary much in tenacity at different temperatures; as might be anticipated *à priori*, those metals which melt most easily become most weakened on heating. Thus Wertheim (*loc. cit.*) found the following values for various annealed wires (as before in kilogrammes per wire of one square millimetre section).

	At 15°.	At 100°.	At 200°.
Iron	46·9	51·1	46·9
Copper	30·5	22·1	—
Platinum	23·5	22·6	19·7
Silver	16·0	14·0	14·0
Zinc	14·4	12·2	7·3
Gold	10·1	12·6	12·1
Cadmium	4·8	2·6	—
Lead	1·8	0·5	—
Tin	1·7	0·85	—

Thus lead, tin, cadmium and zinc, which melt the lowest of the above, are only from one-half to one-third as strong at 200° as at 15°; copper, silver, and platinum are much less weakened on heating; whilst iron and gold are actually stronger at 100° than at 15°, and are but little weaker at 200° than at 100°.

78. Certain alloys possess much greater strength than their constituents, and on this property depends much of their practical use; thus gun-metal, standard gold (gold copper alloy), a silver platinum alloy used for electrical purposes, steel, and phosphor bronze (which last two may be regarded as analogous to alloys) are much more tenacious than either of their constituents severally: thus Matthiessen found the following values (as before in lbs. per double wire, hard drawn, No. 23 gauge).

Metals separately.		*Alloys.*	
Copper	25—30	Gun-metal containing	
Tin	under 7	12 per cent. tin	80—90
Copper. . . .	25—30	Standard gold : gold copper alloy (22	
Gold	20—25	carat gold) . . .	70—75
Silver	45—50	Silver platinum alloy	
Platinum . . .	45—50	(⅔ silver, ⅓ platinum)	75—80
Iron.	80—90	Steel . . .	above 200

As a general rule, however, the effect of alloying metals together is to impair their tenacity; thus gold is rendered brittle by the presence of a trace of antimony, and an alloy of two parts tin and one of platinum is quite brittle.

As already stated, the union of non-metals with metals usually wholly destroys the metallic characters, and in particular this chemical union usually gives rise to a product possessing but little toughness and tenacity such as is requisite for manufacturing purposes (such products as glass, earthenware, bricks, and various minerals, &c., excepted). Steel and phosphor bronze, however, as above stated, form notable exceptions to this rule, the presence of one per cent. or less of carbon in the first, and of a like quantity of phosphorus in the second, communicating to the substances iron and bronze a considerably greater power of resisting wear and tear and other special qualities.

The accurate determination of the tensile strength of wire ropes, bars, &c., of their power of resisting crushing and transverse strains, bending and twisting agencies, &c., is of very great importance to the architect and engineer. Ingenious and powerful machinery for this purpose has been constructed by Mr. David Kirkaldy,[1] whose machines will measure any kind of strain or stress from 10 to 1,000,000 lbs. applied not only to metals, but to wood and building materials generally. Modifications of this testing machinery are employed in some iron-works, &c. for the purpose of continually examining the strength and value of certain of the products (*e.g.* Bessemer rails and the like).

79. **Other physical properties.**—Closely connected with the physical structure which enables metals to exhibit the phenomena of crystallization, malleability, and ductility is the power which some possess of returning to their original shape when deflected therefrom by some external force not too great (*elasticity*); a property possessed to an extreme degree by good steel. The operations of wire-drawing, rolling, hammering, and the like generally increase the elasticity of metals, whilst annealing and fusing usually diminish it. Some metals are almost wholly devoid of elasticity; thus lead scarcely exhibits a trace of this property, being so soft that it is

---

[1] Testing and Experimental Works, 99, Southwark Street, S.E. The specimens in Mr. Kirkaldy's museum, illustrating the results arrived at with many years employment of his testing machinery on all sorts of material, are of a most interesting and instructive character.

readily abraded by the nail. Some metals and alloys, when worked into appropriate shapes and struck, continue vibrating for some time, and hence are powerfully *sonorous* (*e.g.* aluminium, bell metal, steel, standard gold, &c.).

The chief value of many metals and alloys for industrial purposes lies in their possession to a greater or less extent of a combination of properties of somewhat opposite kinds; whilst they possess sufficient rigidity to keep their shape even with moderately hard usage and to bear "wear and tear," when once fashioned into articles of domestic and everyday use, they have the power of yielding to pressure, &c. to a sufficient extent to enable them to be readily worked into these forms. In some cases the requisite softness for this latter purpose is hardly attained until the temperature is considerably raised; thus most articles of wrought iron are made when the metal is softened by heat so as to yield readily to percussion (*forging*) and other shaping processes. Closely connected with this softening or incipient conversion into a pliable mass by heat, is the phenomenon of *welding*, or the adherence together of two separate metallic masses when united by pressure in such a way as to form a join as strong as the other parts. Iron and platinum possess this power at a high temperature; sodium and some of the rarer metals at the ordinary temperature; gold also can be welded cold, under certain conditions, as in gold-beating (§ 72). On the possession of these properties depend most of the metal-fashioning crafts, those where the metals are fused and cast (§ 90) being the main exceptions.

80. Thus in the manufacture of steel pens, as carried out by Messrs. Gillott & Sons, there are no less than eighteen stages between the conditions of bar steel and finished pen; and most of the stages are different appli-

cations of these properties of metals in reference to the shaping of the material into the required form. The bar steel is first converted into thin sheets, which are again rolled to the requisite degree of thinness; from the rolled steel "blanks" are punched out by a machine, leaving a kind of skeleton or network of "scrap steel" (Fig. 26), which is melted up or welded together and used over again. Two "side slits" are then made in the blank (No. 2), and

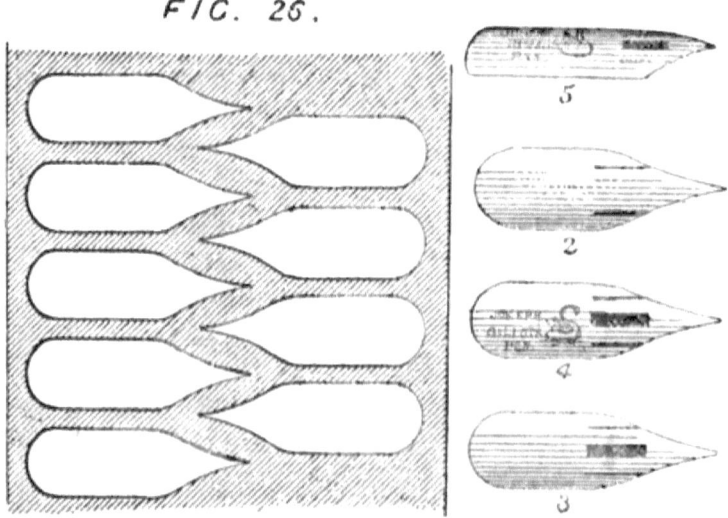

FIG. 26.

a somewhat wider centre slit (No. 3) pierced, a portion of metal being punched out in making this orifice; the metal is then annealed and marked with the maker's name; a device or trade mark is raised by embossing (No. 4), and then the hitherto flat pen is converted into a portion of a cylinder, or curved (technically, "raised") by a suitable machine (No. 5); after which it is hardened, tempered, and cleaned by scouring with emery, &c.; the tip is then "straight-ground," *i.e.* the metal is thinned at the writing end by grinding in the direction of the length of the pen, after which it is "cross-ground," in the transverse direction. Finally the slit from the nib to the punched-out central part is cut, and the pen is coloured and varnished for sale.

Similarly the production of an ordinary pin necessitates a number of stages in the shaping process, which is thus carried out by Messrs. Taylor of Birmingham. Virgin brass (§ 74) having been cast into thin elongated bars, is drawn into wire of the required diameter, which is supplied to the pin-making machine from a drum on which it is coiled. The end of the wire is adjusted to the machine, which gradually draws it through a series of pegs so arranged as to straighten the wire and take the curve from the drum out of it; the wire next passes into a kind of die, when a rapidly acting hammer strikes the slightly projecting end in such a fashion as to flatten it out and fit it into an expansion in the end of the die groove forming the "solid head" (formerly the head was made of a coil of thinner wire slipped over the pin wire, and hammered into shape; such heads were liable to come off). The head being completed, a knife cuts off the proper length from the wire, and the partially made pin is detached from the die and falls into a groove in which it is suspended by the head whilst the other end is abraded to a point by a kind of revolving cylindrical file. All these operations are performed automatically by the machine, and so rapidly that 200 complete pins are turned out of the machine per minute. Finally the pins are cleansed by agitation in barrels with fuller's earth and water, washed, and coated with tin by boiling with granulated tin and cream of tartar, &c.

81. Again, the manufacture of table-spoons and forks, many kinds of brass-work, cutlery, percussion-caps, copper

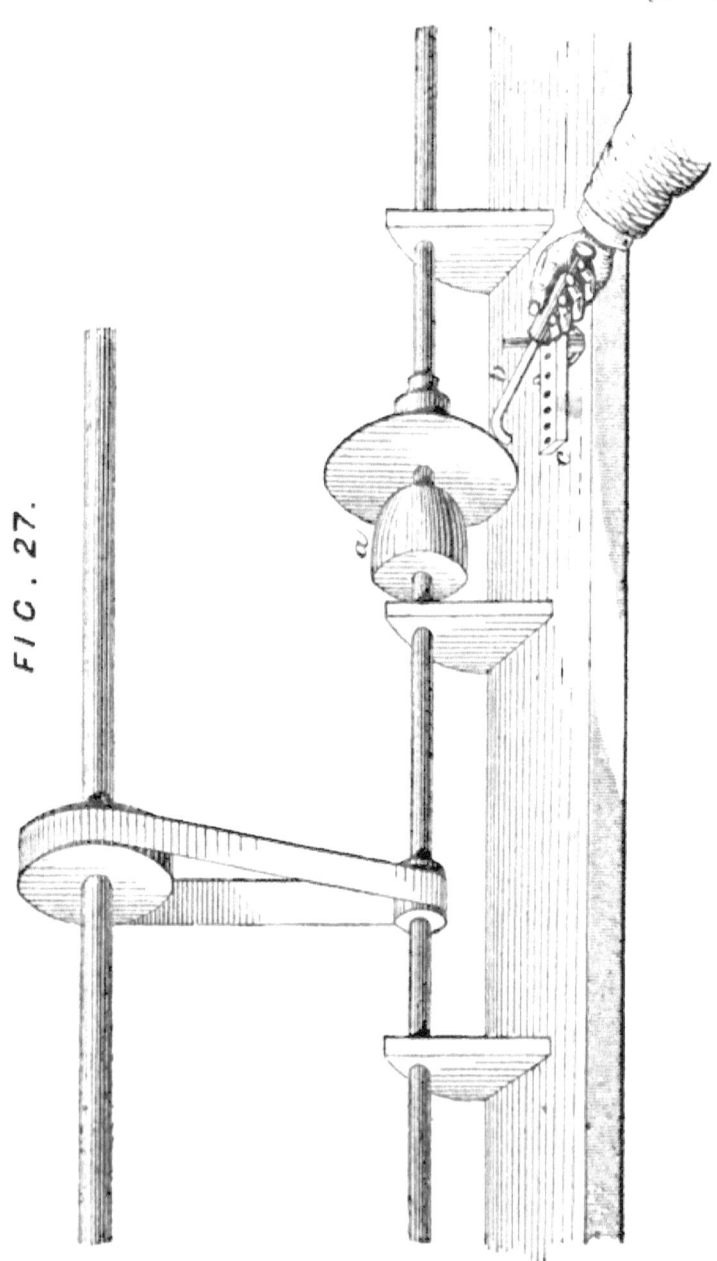

FIG. 27.

pans and kettles, medals and coins, and a thousand-and-one articles of every-day use, all depend upon the possibility of forcing the metal into various shapes without fracturing it, by mechanical processes, such as forging, punching, pressing, embossing, and the like. One of the prettiest illustrations of the application of pressing and shaping force is afforded by the processes in use for "teapot spinning," *i.e.* the production of a Britannia-metal teapot by a process technically termed *spinning*. The alloy (§ 103) being rolled into sheets of convenient thickness, a circular disc is cut out and placed in a kind of lathe as represented in Fig. 27, the metal disc being

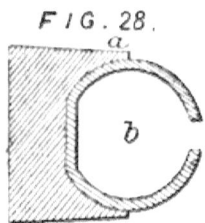

FIG. 28.

pressed against a nearly hemispherical wooden chuck *a*. The lathe being set in motion, the workman presses against the off-side of the disc with a peculiarly shaped tool, *b*, held steadily by means of the rest, *c*, so as gradually to bend the disc over the mould, *a*, and so to convert the disc into a bowl. The bowl thus formed is taken off the lathe and set with the convex part fixed into the concavity of a hollowed-out chuck (shown in section *a*, Fig. 28); by the aid of two differently shaped tools held one in each hand and applied, the one within and the other without the rim of the bowl, the metal is gradually bent inwards as it revolves, so as finally to take an almost globular shape: Fig. 29 indicates the closing stage of this operation, the nearly globular bowl thus

L

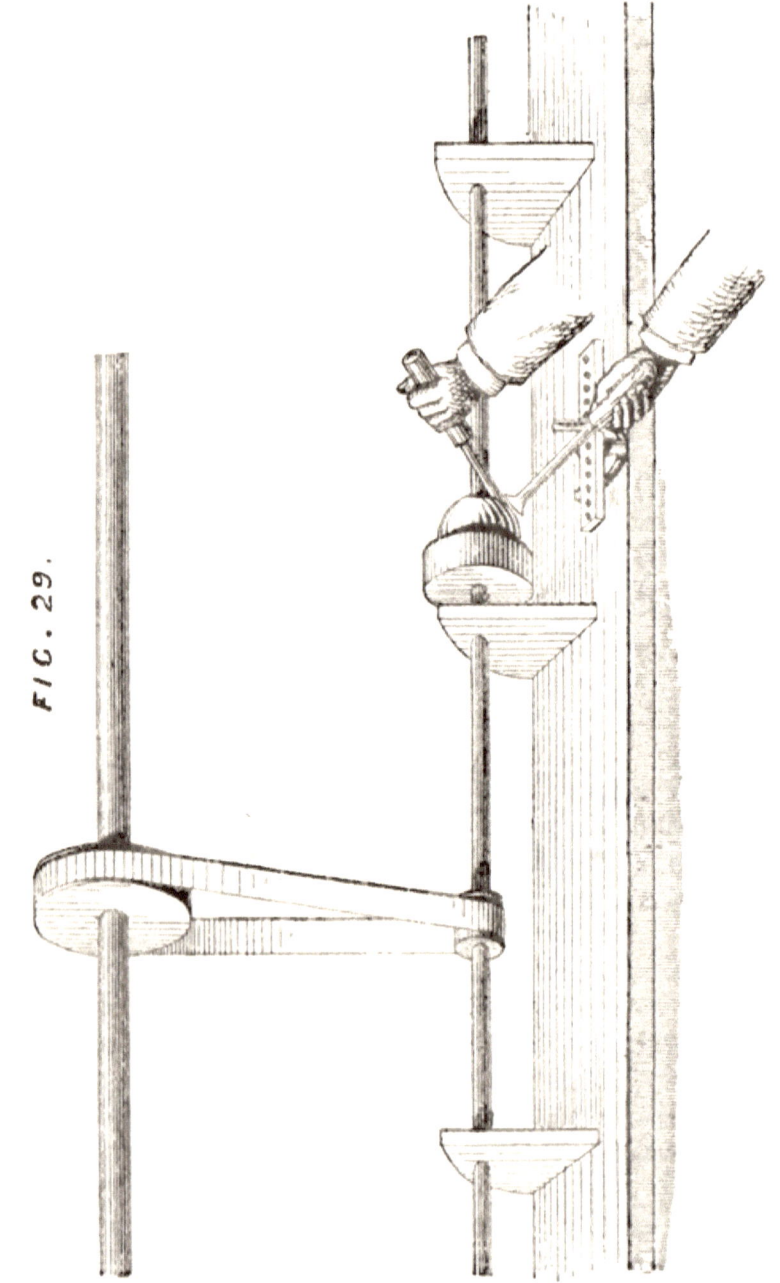

FIG. 29.

formed being shown in section in Fig. 28, *b*. Finally the lid, spout, handle, &c., are attached, and the whole brightened and polished for the market. During the spinning the edge of the disc, some forty or fifty inches in circumference, becomes diminished to almost half that in the bowl, and to about one quarter in the globular pot, the metal being thus as it were pressed in upon itself, as well as somewhat extended, the superficial area of the outside of the globular pot being somewhat greater than that of one side of the circular disc used in the first instance. In a similar fashion jugs and analogous vessels are "spun up," out of plates, the lips for pouring being subsequently shaped by carefully hammering or pressing out the metal on a wooden or metal mould. Silver articles, *e.g.* bowls, teapots, &c., are frequently curved by an analogous operation; the second stage, however, cannot so well be applied to silver, so that if a closed-in vessel is required like a teapot, it is usually made in two halves, neatly soldered together.

CHAPTER VI.

THERMIC AND ELECTRIC RELATIONS OF METALS.

82. WHEN heat is applied to a metal or alloy, the substance increases in temperature, the heat passing from the parts directly heated to the rest of the mass by the process of *conduction:* simultaneously the whole increases in size (*expansion*) and ultimately softens and liquefies : if the heat be kept high enough, volatilisation to a greater or less extent usually takes place, this phenomenon sometimes occurring without previous fusion, as in the case of arsenic.

Metals possess the power of conduction to very different extents; thus Wiedemann and Franz give the following values :—

Conductivity of Metals for Heat.

Silver	100·0
Copper	73·5
Gold	53·2
Tin	14·5
Iron	11·9
Lead	8·5
Platinum	8·4
Bismuth	1·8

Amongst the practical applications made of the high power of conducting heat possessed by some of the metals may be instanced the Davy lamp, in which ignition of an explosive mixture of fire-damp and air outside the lamp by the flame inside is prevented (with due care), inasmuch as the temperature produced by the burning of the inflammable air inside the lamp is so far reduced by the wire-gauze cover conducting away the heat, that inflammation of the gaseous mixture through the gauze does not take place, the temperature just outside not reaching that required to ignite the explosive mixture.

The conductivities of alloys are in some few cases almost exactly those that may be calculated from the conductivities of the constituents on the assumption (as in the case of specific gravity) that each metal exists in the alloy side by side without mutual interference; so that an alloy of this kind composed of equal volumes of any two metals will possess a conductivity equal to the arithmetical means of the conducting powers of the two constituents. In other cases, however, a wide difference exists between the conductivity thus calculated and that observed. According to Matthiessen all the metals except lead, tin, zinc, and cadmium, when alloyed together, or with one of these four, so as to form a binary alloy, yield a product of which the conductivity is uniformly more or less below that calculated on the above assumption; whilst any pair out of these four yield a binary alloy the conductivity of which agrees very closely with that calculated.

83. **Conductivity for Electricity.** — It is worth noticing in passing that the conductivities of metals for electricity are in many cases closely allied to those for heat; there is a high probability that these two powers are really identical, and that in those cases where a discrepancy exists, the cause of this is a different or

imperfect state of purity in one or other of the substances examined. Thus the following table gives the results of most carefully made experiments by Matthiessen on the electric conductivities of various metals in a high state of purity: and on contrasting these with the values given by Wiedemann and Franz for heat conductivities (§ 82), it is at once visible that, save in the cases of copper, gold, and platinum, the two series are practically identical. These three exceptional metals happen to be greatly influenced in conductivity by the presence of traces of impurity: moreover the relative values obtained at one temperature are not quite the same as those at another, the electric conductivities of metals being diminished by rise of temperature, but not always at the same rate in each case.

Silver	. . .	100·00	Iron	. . . .	16·81
Copper	. . .	62·95	Tin	. . . .	12·36
Gold	. . . .	77·96	Lead	. . . .	8·32
Aluminium	. .	56·06	Antimony	. . .	4·62
Zinc	. . . .	29·02	Bismuth	. . .	1·24
Platinum	. .	18·03			

A pretty illustration of the different conducting powers for electricity exhibited by different metals is afforded by passing a tolerably powerful current through a chain made of alternate links of silver and platinum wire; the latter being the worse conductor (*i.e.* offering more *Resistance*) becomes much more heated, so that the links are heated alternately to dull redness or somewhat below, and to bright incandescence.

84. The conducting power of a given wire varies considerably according as the wire is hard-drawn, or has been annealed; as a rule annealing improves conductivity. The influence of small quantities of impurities in diminishing conductivity is probably due largely to the

hardening effect thus produced; thus 0·2 per cent. of iron in a copper wire diminishes the conductivity to only three-quarters of the original amount, and a trace of arsenic to one-third: on the other hand, the effect of heating a metal is to diminish its conductivity, although it might be supposed that the effect of a rise of temperature would be a *quasi*-softening. Most metals lose about 30 per cent. of their conductivity at 0° on heating to 100°, the loss being slightly different in each case: iron loses 38 per cent. This alteration in conductivity on heating has been utilized by Siemens in the construction of a valuable pyrometer, which may be briefly described as a coil of platinum wire suitably protected and attached to a hollow iron pole, so that the end where the coil is can be thrust into the furnace, &c., the temperature of which is to be measured. A current of electricity is led, by insulated wires passing through the hollow pole, to the coil, the current forking before reaching the coil, so that there are two return currents, one passing through the coil, the other not. By making this latter pass through a known resistance, and then measuring the relative currents passing through the coil and through the known resistance, the resistance of the coil when heated is ascertained; and as its original resistance is known, and the rate at which platinum alters in resistance with the temperature is known, the temperature to which the platinum coil is heated can be readily calculated. In the later forms of the instrument the relative strengths of the two currents are determined by enclosing in the two branches of the circuit two ingeniously-contrived voltameters, so that the columns of gas produced after a few minutes' action can be read off; from these numbers, the temperature of the coil is known by reference to a specially constructed table.

The conductivities for electricity of alloys appear to follow the same laws as those for heat, the four metals lead, tin, zinc, and cadmium constituting a class separate from the others (§ 82.) The effect of temperature on the conductivity of an alloy is in some cases much the same as upon a simple metal, a rise of temperature from 0° to 100° diminishing the conducting power from 25 to 30 per cent. Other alloys suffer much less diminution— thus, German silver loses only about 4 per cent., and a silver-platinum alloy of two parts silver and one platinum only 3·1 per cent. ; hence these alloys are valuable for the preparation of standards of electrical resistance. Non-metallic substances as a rule are *increased* in conducting power by heat.

85. **Specific Heat.**—In order to raise the temperature of equal masses of various metals through the same range, very different amounts of heat are requisite : Dulong and Petit have shown that the amounts of heat required to raise the temperature of 1 gramme of any metal from 0° to 1° (the "specific heats" of the metals respectively) are very nearly inversely proportional to the "combining numbers" of these metals as deduced from their chemical behaviour ; the actual values of the specific heats are such that the following equation always holds approximately,

$$S \times C = 6\cdot3,$$

when S is the specific heat and C the combining number of any given metal. This rule, moreover, is not confined to metals, all non-metals that are solid at the ordinary temperatures being found also to conform to it with three exceptions, viz., carbon, silicon, and boron ; whilst from the recent researches of Weber and others, it seems that although these bodies are exceptional

VI.]   CHIEF INDUSTRIAL APPLICATIONS.   153

when the specific heats at the ordinary temperatures are taken, they conform to the rule at high temperatures.

Thus the following table exhibits the relationship between the combining numbers and specific heats of the more important metals :—

Metals.	Combining Number.	Specific Heat (Regnault).	Product.
Aluminium	27	0·2143	5·8
Antimony	122	0·0508	6·1
Arsenic	75	0·0814	6·1
Bismuth	210	0·0308	6·5
Cadmium	112	0·0567	6·3
Copper	63·5	0·0951	6·0
Gold	196	0·0324	6·4
Lead	207	0·0314	6·4
Iron	56	0·1138	6·1
Magnesium	24	0·2499	6·0
Manganese	55	0·1217	6·7
Mercury (solid)	200	0·0325	6·5
Nickel	59	0·1089	6·4
Palladium	106	0·0593	6·3
Platinum	197·6	0·0329	6·5
Potassium	39·1	0·1695	6·5
Silver	108	0·0570	6·2
Sodium	23	0·2934	6·7
Tin	118	0·0562	6·6
Zinc	65	0·0956	6·2

This relationship may be illustrated in the following way : two blocks of lead and zinc are prepared weighing respectively 207 and 65 grammes; these are immersed in water kept boiling for some minutes so as to acquire the temperature of 100°, and are then lifted out and quickly transferred to two similar vessels each containing 500 ccs. of water: two beakers answer well, the blocks of metal being gently lifted and lowered in by means of pieces of string or thin wire attached to them. After two or three minutes the water in each beaker is stirred up well to make it acquire a uniform temperature, when it is found that each possesses sensibly the same temperature, so that a Matthiessen's differential thermometer (a differential air-thermometer with two pendent

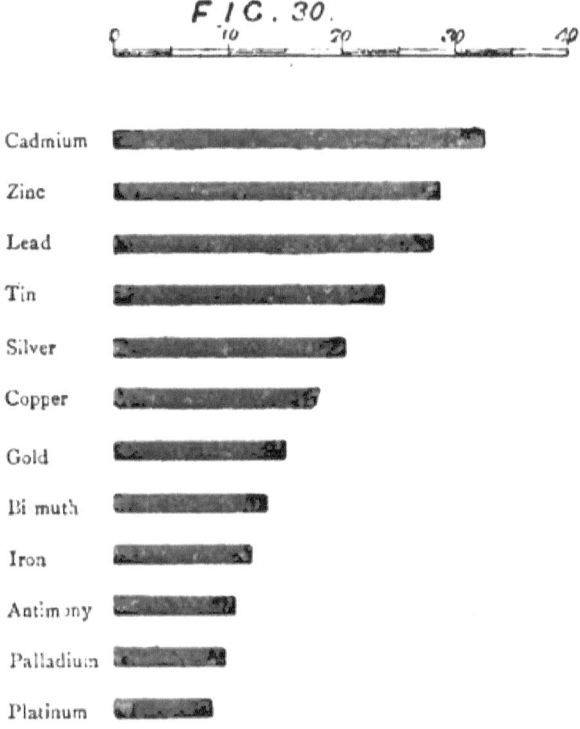

bulbs) exhibits no motion of the index column of liquid when one bulb is placed in one and the other in the other beaker; *i.e.* the same amount of heat is communicated to the water by 65 grammes of zinc as by 207 grammes of lead.

When metals are alloyed together, the specific heat of the alloy is very close to that calculated from the composition of the alloy and the specific heats of the components, the relations of alloys generally to specific heat and specific gravity being alike in these respects (§ 69).

86. **Expansibility.**—The rate at which metals increase in bulk as the temperature rises is usually nearly proportionate to the increase in temperature, though not absolutely so. Slight differences in purity appear greatly to affect the co-efficient of expansion, somewhat considerable differences in the numbers obtained by different experimenters being occasionally noticeable. The following table illustrates the average numerical values of the increments in length of bars of different metals found by various observers on heating from 0° to 100°; the expansions in area being almost exactly double, and those in volume treble, these numbers. Fig. 30 indicates the same values drawn graphically to scale, the original length of the bars being 500 inches (or 303 times that of the longest line).

Increase in length of 10,000 *units on heating from* 0° *to* 100°.

Cadmium	. . .	33	Gold	. . .	15
Zinc	. . . .	29	Bismuth	. . .	14
Lead	. . . .	28·5	Iron	. . . .	12
Tin	. . . .	24	Antimony	. . .	11
Silver	. . . .	20	Palladium	. . .	10
Copper	. . . .	18	Platinum	. . .	9

The different expansibility of metals is well illustrated by taking a compound bar of two metals riveted together; if the bar be flat to start with, on heating it becomes sensibly curved, the most expansible metal being outside. This property is made use of in the construction of Martin's compensating pendulum, where the lengthening of the rod of the pendulum on increase in temperature is just compensated by the rise in the position of the centre of oscillation of the whole pendulum, by the raising of weights fixed at the ends of a compound bar traversing horizontally the pendulum rod, and fixed thereto at its centre, the most expansible metal being placed downwards. Similarly *Breguet's thermometer* acts by the twisting or untwisting of a spiral composed of two or more metals of different expansibility, an index being moved over a dial by the motion of one end of the spiral, the other being held fast by a clamp: *Harrison's gridiron pendulum*, where the expansions of different metals are ingeniously contrived so as to counteract one another, and *Graham's mercurial pendulum* (a reservoir of mercury at the end of a steel rod), similarly, are practical applications of the different expansibilities of different metals, whilst the mercurial thermometer and several forms of pyrometers for the measurement of high temperature all depend on the expansion of mercury or other metals.

87. In the act of expansion an enormous amount of force is exerted; conversely, if the ends of a heated bar of metal be fixed to two obstacles, and the bar be allowed to cool, either the obstacles are forced nearer together or the bar becomes permanently extended. To elongate a bar of iron one square inch in section by $\frac{1}{10,000}$ of its length requires a tensile force of one ton (Barlow): now if a bar of these dimensions be heated to 100° and then allowed to cool to 0°, it diminishes in length $\frac{12}{10,000}$ of its length; and hence, to prevent it contracting, a force of not less than twelve tons must be applied. Thus a rod of cast-iron is readily broken if passed through a perforation in a hot iron

bar fixed in a frame so that when the bar shrinks on cooling the centre of the rod is forcibly pulled inwards, the ends of the rod being prevented from moving by being fixed against two projecting parts of the frame.

Great care has accordingly to be used when metal girders are employed for constructive purposes, &c., as, if due space be not left for expansion and contraction, the bars forcibly thrust or pull out of their proper positions the walls, &c., into which they are built: on the other hand, if a rod of iron be passed through a building, and plates fixed to its projecting ends so as to press against the wall when the rod is expanded by heating, the walls will be forcibly drawn inwards during contraction on cooling, and thus bulging walls may be gradually restored to the perpendicular. In this way the *Conservatoire des Arts et Métiers* in Paris was preserved from falling, as have several other buildings subsequently. The tires of wheels are put on hot, and the rivets of boiler-plates, &c., are applied hot, so as to bind together more firmly the parts to be united through the shrinkage on cooling.

Alloys usually expand at about the same rate as would a compound bar of the same length composed of the constituent metals placed end to end so as to be present in the same proportions as in the alloy.

88. **Fusibility.**—Most metals when heated sufficiently pass into the liquid state; but the temperature required to produce this phenomenon varies much: the following table illustrates the different fusion-points of various metals. Different experimenters have given considerably different figures in various cases, the discrepancies being doubtless due to imperfect purity of the bodies examined, and to the circumstance that it is not easy to measure elevated temperatures very exactly.

	Deg.		Deg.
Mercury	− 39	Zinc	near 420
Potassium	+ 62	Antimony	,, 450
Sodium	97	Silver	,, 1020
Tin	near 230	Copper	,, 1090
Bismuth	,, 270	Gold	,, 1100
Cadmium	,, 320	Iron	,, 1500
Lead	,, 330	Platinum	,, 2000

Mercury is the only metal fluid at ordinary temperature; next to it in fusibility comes the newly discovered element gallium, which melts at a little above 30° (Lecoq de Boisbaudran).

Alloys almost invariably melt at temperatures considerably below those calculated from the quantities of metals present and their respective fusing-points; this property leads to the practical employment of many such alloys for casting, soldering, &c.

In illustration of this property it may be noticed that a mixture of potassium and sodium is fluid at the ordinary temperature. An alloy of three parts lead and one tin (melting respectively at near 330° and near 230°) fuses at 283°, the calculated melting-point being $\frac{3 \times 330 + 230}{4}$ or 305°; whilst one of rather more than seven parts lead to twelve of tin (corresponding to the formula Sn_3Pb) melts at 181° (Pillichody), the calculated melting-point being near 265°. On addition of bismuth to a lead-tin alloy, the fusing-point is still further lowered; thus the ternary alloy known as "fusible metal" (lead five parts; tin three; bismuth eight) melts at 100°, whilst the similar alloy (lead two parts; tin three; bismuth five) melts at 93°. On further adding cadmium to a bismuth-lead-tin alloy a still further reduction in melting-point ensues; thus "Wood's alloy" (bismuth fifteen parts; lead eight; tin four; cadmium three) melts completely at 63° C, becoming soft at 57°. Such alloys therefore readily melt in the steam of boiling water, and the last in the vapour of boiling absolute alcohol. The use of *solders* of various kinds for uniting metal-work depends on the low melting points of the various alloys used ; on applying a certain degree of heat, insufficient to fuse the metalwork itself, the soldering applied melts and adheres to the surfaces like quicksilver; on cooling, the solder solidifies and thus permanently fastens together the metals, much as papers are pasted together.

It is noteworthy that in cases where a large amount of heat is liberated during the intermixture or combination of the constituent metals in an alloy, the melting-point of the alloy is not always lower than that calculated from the quantities of the constituents and their melting-points. Thus the addition of a few per cents of sodium (melting at $+ 97°$) to mercury (melting at $- 39°$) gives an alloy or amalgam solid at the ordinary temperature, and therefore melting far above the calculated point.

89. Although by incorporating metals together in a state of fusion a tolerably intimate mixture is produced, yet it sometimes happens that on standing in a fluid state for some time, more or less separation of the two bodies occurs ; or rather, two alloys separate from one another, differing in composition, one metal usually predominating in one and the other in the other. Thus if equal weights of bismuth and zinc be mixed in the melted state and cast in a red-hot ingot mould which is maintained at a high temperature for some little time so as to allow of separation taking place during cooling, the lower part of the ingot will consist of a bismuth-zinc alloy containing 8·14 per cent. of zinc, whilst the upper one will be a zinc-bismuth alloy containing 2·4 per cent. of bismuth (Matthiessen). The same kind of thing

happens with lead and zinc, the heavier alloy containing 1·6 per cent. of zinc, and the lighter 1·2 per cent. of lead. This result is precisely analogous to what takes place on shaking anhydrous ether and water together; on standing, water falls to the bottom, containing a small amount of ether dissolved, whilst ether floats to the top, retaining a minute quantity of water in solution. A similar result is obtained if fusel oil (coloured red by a little magenta) be shaken up with chloride of copper solution (green) : this mixture is neutral in colour, but speedily separates again in a red and green layer.

This tendency towards separation is peculiarly inconvenient in various kinds of castings, and has to be overcome by special devices. The grain of grey cast-iron is partly due to a partial separation of the dissolved carbon during cooling (§ 35); bell-metal is very liable to separate more or less, so that different portions of a large bell may have perceptibly different compositions; the art of the bell-founder partly consists in adding together the constituents in such proportions, and so manipulating, as to prevent as far as possible this tendency to separate. The standard silver of Great Britain (silver-copper alloy containing silver 92·5, copper 7·5 per cent.) has a tendency to separate in this way, the outsides of cast bars of this alloy being slightly poorer in silver than the centres; silver-copper alloy containing 71·9 per cent. of silver has, however, no tendency to separate; whilst alloys containing still less silver separate somewhat, but in these cases the outsides of the bars become the richest in silver.

This tendency is occasionally utilized to effect the partial separation of metals from one another, especially if there be a decided difference in fusibility; thus Pattinson's lead desilverizing process depends on this principle the alloy richest in lead being the most fusible; again the process termed "eliquation," or "liquation," is another case in point: to extract small quantities of silver from copper, the whole is melted with lead and the mass heated; a lead-silver alloy then separates from a copper-lead alloy, the former running out in a fused state, the latter remaining as a spongy skeleton of the mass (§ 20).

90. In order to prepare a "casting," it is necessary that hollow moulds should be constructed, into which the molten metal is poured from a crucible or melting-pot, or allowed to run from the melting furnace by a gutter according to the size of the object to be cast. According to the shape of the object the construction of these moulds varies; if small solid castings are to be made, and the metal fuses readily, the moulds are made of metal, plaster of Paris, or other similar material, being composed of two halves which when put together will enclose a cavity (e.g. a pistol-bullet mould): if the casting is of any size, *two* holes are constructed in the mould, one to pour the metal in, and the other to allow the enclosed air, &c. to issue freely. For objects of cast-iron, gun-metal, and the like, moulds of clay and loam, sandy earth, &c., are employed, and, if the object be tubular or hollow, are so constructed that a central *core* is erected on the floor of the founding-pit, and round this a larger hollow mould is set so as to leave a space between the core and the outer mould; the metal being then run into this space, the required object is obtained. Thus, in founding a large bell, a hollow core is built up of brickwork, *a a a*, Fig. 31, and plastered over with clay, *b b b;* the soft clay surface is then turned to the exact shape of the inner surface of the bell by cutting away the superfluous clay by

M

means of a "crook," *c*, consisting of a wooden board or metal sheet held by an arm attached to a vertical axle, *e*, which is passed through the hollow part of the core to a bearing on the floor to give steadiness; as the shaft revolves, the clay is cut away to the required extent by the revolving crook. To the same arm is fixed

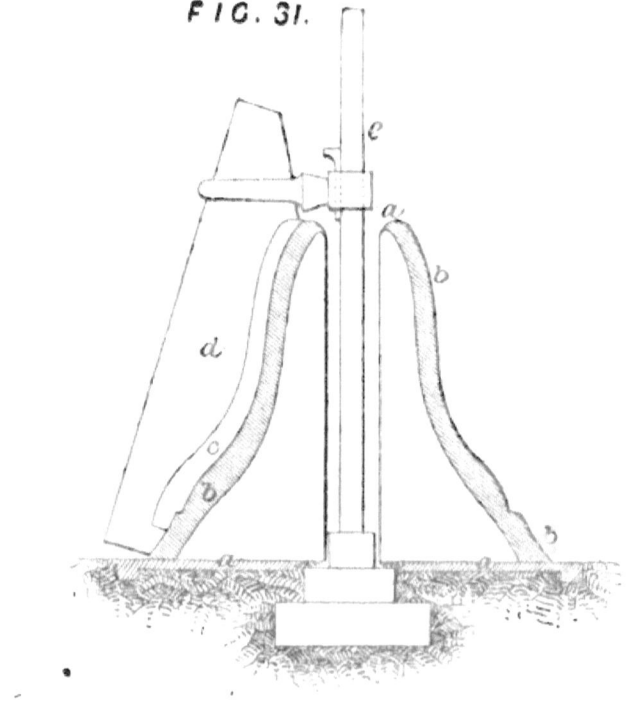

FIG. 31.

a similar cutter, *d*, shaped exactly to the *outside* section of the intended bell; when the core is shaped it is dried by kindling a fire in the hollow part, and is then covered with a greasy composition (to prevent adherence); haybands are then twisted round it, and clay spread over the covering; the cutter *c*, of the crook being removed, the clay is now cut away by the outer cutter, d, so as to give the exact shape of the outer bell surface. This coating is then dried as before, and on it the clay forming the outside mould or "cope" is plastered to a sufficient thickness and dried. The

FIG. 32

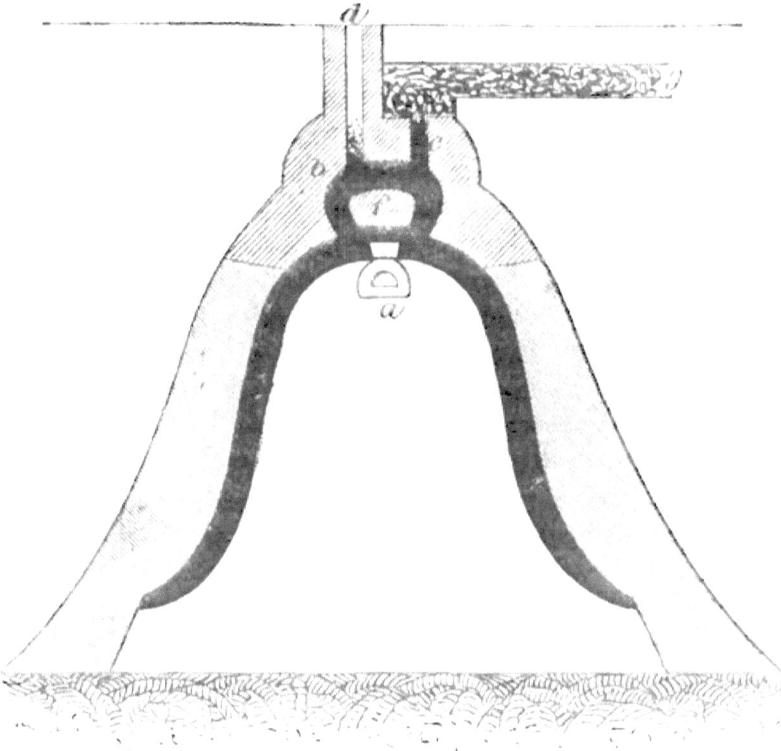

cope being removed, the haybands, &c. are stripped from the core, which is then placed on the floor of the casting-bed, and covered with the cope in the exact position to enclose the bell-space between the two surfaces, Fig. 32. A staple, a, to hold the clapper is placed in the top of the core, and a separate mould, *b*, to form the head of the bell or *crown*, affixed to the top of the cope; two holes, *c* and *d*, pass upward from the "ear," or "cannon," *f*, (the projection by means of which the bell is hung). Loam is then shovelled into the pit and rammed down all round the cope until the pit is full up to the level of the top of the upper mould; a gutter, *g*, is then traced on the surface of the loam from the tapping hole of the furnace to one of the holes leading to the internal cavity, and the metal run in, the displaced air passing out at the other hole. An ordinary-sized bell takes twenty-four hours to cool down; but one of the size of "Big Ben" of Westminster ($13\frac{1}{2}$ tons) would not be sufficiently cool to dig out for four days or more. When the casting is cool enough the moulds are destroyed by digging, and the bell withdrawn from the pit and trimmed and tuned by hand; the tuning being effected by cutting away by machinery a small part of the substance so as to thin the metal at the edge or above where the clapper strikes (the *soundbow*) or elsewhere, according as the tone is to be raised or lessened a little: for a peal of bells a close approximation to the exact tone is obtained by properly shaping the core and cope in the first instance, *i.e.* by suitably adjusting the sizes and shapes of the two cutters of the "crook," that regulate the outer and inner bell-surfaces.

91. In order to produce "sharp" castings, it is essential that the metal or alloy used should occupy when just melted slightly *less* space than when solid and just upon the melting point; so that as it solidifies it may thrust itself into all the crevices of the mould and take a fine impression. With most metals a shrinking takes place on solidifying, wherefore they cannot be

used for objects in which sharp impressions are required : thus a counterfeit sovereign or shilling prepared by casting in a mould is generally deficient in sharpness of impression. Zinc is, however, largely used for the manufacture of small cheap articles for ornamental purposes, notwithstanding this defect. Fusible metal, type metal, bismuth, antimony, cast-iron, brasses, and bronzes, and some few other alloys, (§ 103) are the most suitable for founding and casting on account of their greater or less expansion during solidification ; type metal indeed derives its name from its use in casting printing types and stereotypes, the cold alloy being hard enough to resist to a considerable extent the pressure of the printing press, and general wear and tear.

When large castings are made, required of certain dimensions, it is necessary to make an allowance for shrinkage during cooling after solidification whilst preparing the moulds; thus with cast-iron it is usual to make the moulds about 1 per cent. larger linearly than the casting is required to be, a shrinkage of one-eighth of an inch to the foot being allowed for. With zinc the shrinkage is greater. Greatly increased strength is given to large castings by applying hydraulic pressure during the operation, as proposed by Sir Joseph Whitworth; in this way the formation of air-bubbles and cavities (*honeycombing*) is largely prevented.

92. **Annealing, hardening, and tempering.** —The effects of heating and allowing to cool slowly (annealing) on the malleability, tenacity, and other physical properties of metals have been referred to in Chap. V., whilst the hardening of steel produced by almost instantaneous cooling, and the "tempering" of hardened steel by a gentle annealing have been described in § 41. It is noteworthy that bronze is differently

affected by sudden chilling when red hot; instead of being hardened it becomes softer and more malleable.

93. **Volatility.**—Probably all metals can be volatilized at a very elevated temperature; some, however, are much more readily converted into vapour than others. Arsenic volatilizes without previous melting, subliming in a lustrous mirror when the operation is conducted in a glass vessel, especially if filled with a gas which will not act chemically on the metallic vapour. Mercury boils at near 350°, cadmium at about 860°, zinc at 1040° (Deville), whilst magnesium can be distilled in a current of hydrogen at a white heat, and silver in the oxyhydrogen flame (Stas); gold and copper, and particularly lead, are sensibly volatilized during melting; potassium and sodium are volatilized and condensed during the process of manufacture at nearly a white-heat (§ 52).

When alloys of a readily volatile and a non-volatile, or difficultly volatile metal are heated sufficiently, the volatile metal is expelled. Thus a solution of potassium, sodium, lead, gold, &c., in mercury (termed an *amalgam* of potassium, &c.) loses its mercury on heating, leaving behind the other metals (§§ 16 and 22). This property is occasionally utilized in metallurgical processes, as in water-gilding and silvering (§ 101). Platinum is sometimes alloyed with arsenic, and the alloy fashioned into the required shape; the article is then intensely ignited for some time, when the arsenic is expelled. In making brass castings, &c., special precautions have to be adopted to avoid loss of the volatile constituent zinc.

94. **Thermo-electricity.**—When two bars of dissimilar metals are united together, and the junction heated, a disturbance of the electric equilibrium ensues, so that one metal acquires a higher potential than the

other, a current accordingly flowing through a wire uniting the free ends of the two bars, and passing through the wire from the bar at the higher potential to the lower.

The strength of the current depends on the nature of[1] the metals and on the difference of temperature between the junction and the free ends; it also depends somewhat on the absolute temperature of the junction, as with certain pairs of metals a current is set up in one direction by heating the junction to one temperature, but in the opposite direction if heated to another temperature. The following table gives the relative potentials assumed by various pairs of metals, and some other substances, on heating the junctions to equal extents; the numbers attached give the comparative electromotive forces of the currents by taking their algebraic differences: thus, on heating a copper-bismuth couple a current will pass through the conducting wire from the copper to the bismuth, the electromotive force of the current being

$$-1-(-25) = +24.$$

Similarly, with an antimony-zinc couple the current will pass from the antimony to the zinc, the electromotive force being

$$+9.87 - (+0.21) \text{ or } +9.66,$$

whilst with an antimony-bismuth couple the electromotive force will be

$$+9.87 - (-25) \text{ or } +34.87.$$

---

[1] In several leading text-books on chemistry and chemical physics the direction in which the current flows is misstated as being precisely the opposite to that in which it actually does flow, the errors being repeated in the diagrams given!

	Relative potential.		Relative potential.
Bismuth (cast)	$-25 \cdot$	Platinum	$-0 \cdot 72$
Cobalt	$-9 \cdot$	Silver	0
Potassium	$-5 \cdot 5$	Gas carbon	$+0 \cdot 06$
German silver	$-5 \cdot 2$	Zinc	$+0 \cdot 21$
Nickel	$-5 \cdot 02$	Cadmium	$+0 \cdot 33$
Sodium	$-3 \cdot 09$	Arsenic	$+3 \cdot 83$
Mercury	$-2 \cdot 52$	Iron (pressed wire)	$+5 \cdot 22$
Aluminium	$-1 \cdot 28$	Red phosphorus	$+9 \cdot 60$
Magnesium	$-1 \cdot 17$	Antimony (cast)	$+9 \cdot 87$
Lead (pressed wire)	$-1 \cdot 03$	Tellurion	$+179 \cdot 8$
Copper	$-1 \cdot$	Selenion	$+290 \cdot$
Tin	$-1 \cdot$		

With the exception of gas carbon, the low conductivity for heat of the non-metals given in this table prevents their employment for currents of any considerable duration; the numbers apply only to cases where the temperature-difference between the junction and free ends is the same in each instance; thus, in the above supposed cases (bismuth-copper, zinc-antimony, and bismuth-antimony couples), the electromotive forces will only be as 24 to 9·66 to 34·87, provided the junctions and free ends have the same temperature-difference in each case. Moreover, the above numbers are only true when the junction is heated to a temperature between 4° and 38° C., the numerical values being in some cases different at other temperatures; thus at 0° the current flows through the conducting wire from zinc to silver; at higher temperatures up to 120° the current is in the same direction, and of greater electromotive force; beyond 120°, however, it declines, becomes nil, and finally the current becomes reversed, *i.e.* it passes from silver to zinc through the conducting wire. Within certain limits of temperature, the electromotive force is usually nearly proportionate to the temperature-difference between the two ends of a couple.

95. When a junction is cooled instead of heated, the

CHIEF INDUSTRIAL APPLICATIONS.

opposite effect is produced, *i.e.* the current flows in the opposite direction, the metal which acquired the higher potential when the junction is heated taking the lower potential when it is cooled.

Peltier has found that converse actions are produced when weak currents of electricity are made to flow through junctions of two dissimilar metals. Thus if a current be made to flow through a junction of bismuth and antimony, so that it passes from the bismuth to the antimony through the junction, the temperature of the junction falls; whilst if the current flows the opposite way, the junction becomes heated; that is, whilst heat must be applied to the junction in order to render the antimony of the higher potential, if this latter effect is brought about by extraneous means (as a separate battery), the junction becomes heated; whilst on the other hand artificially cooling the junction causes the antimony to take the lower potential; and, reciprocally, artificially making the antimony of the lower potential causes the junction to become cooled.

The production of electricity by heat is sufficiently extensive to enable powerful thermo-batteries to be used as very efficient substitutes for voltaic cells, although a single thermo-couple yields but little current. Matthiessen found that a bismuth-tellurion couple when heated so that the junction was at $100°$ and the other ends at $0°$ had an electromotive force of about $\frac{1}{25}$ volt; or a battery of 25 such couples would almost equal 1 Daniell's cell, the electromotive force of which is $1·12$ volts. Various forms of thermo-electric batteries for practical use have been proposed; and some (*e.g.*, Clamond's) are practically used by electro-metallurgists for electro-plating; most of them, however, are liable to deteriorate greatly after being in use for a long time, either owing to the oxidation of the metals, or the incipient separation of the metals from the other elements of the pile (where compositions are used), thus diminishing greatly the current.

96. Galvanism, or Voltaic Electricity.—

When two pieces of dissimilar metals (or certain other substances) are immersed in a fluid which acts unequally upon them chemically, a result is brought about analogous to that produced on heating the junction of a thermo-couple; different potentials are assumed by the two metals, and on connecting them by a conductor a continuous current is produced as the chemical action continues; the direction of the current and its relative strength depends (*inter alia*) on the nature of the two metals, and, even when they are the same, on the character of the fluid. Thus Faraday found the following chemico-electric order of various ordinary metals when placed in the fluids specified, the metal highest in the list acquiring the higher potential (*i.e.* the current passing through the connecting circuit from the metal highest in the list to that lower down) :—

In dilute sulphuric acid.	In caustic potash.	In potassium sulphydrate.
Silver.	Silver.	Iron.
Copper.	Nickel.	Nickel.
Antimony.	Copper.	Bismuth.
Bismuth.	Iron.	Lead.
Nickel.	Bismuth.	Silver.
Iron.	Lead.	Antimony.
Lead.	Antimony.	Tin.
Tin.	Cadmium.	Copper.
Cadmium.	Tin.	Zinc.
Zinc.	Zinc.	Cadmium.

One of the most important applications of this principle consists in the use of "galvanic batteries," (series of pairs of plates excited by various chemical fluids) to develop electrical currents for telegraphic purposes; whilst currents thus generated are mainly employed in electrometallurgy, although thermo-currents (§ 95) and currents produced by magnetic induction (§ 98) are frequently substituted for "voltaic" currents (currents generated by differential chemical action on two dissimilar metals). Frequently gas-carbon is substituted for the "electro-negative" metallic plate in such a combination (the one acquiring the higher potential), metals not being the only substances capable of forming voltaic couples; the other element of the couple, however, is in practice almost always zinc.

97. **Magnetism, Electro-magnetism, and Magneto-electricity.**—When an electric current circulates through a conductor capable of free motion, the conductor arranges itself in a defined position with reference to the earth dependent on the direction of the current and the position of the apparatus on the earth's surface; a wire formed into a helix conveying a current will arrange itself (in England) so that the axis of the helix points more or loss north and south, the south end being that where the current circulates in the direction of the hands of a watch, whilst the north end dips below the horizon. The quasi-magnet thus formed has also the power of attracting light particles of soft iron or steel, and of magnetizing the latter or of communicating to it the permanent power of behaving in a similar fashion, *i.e.* of possessing "magnetic polarity." The same power is also possessed by certain natural iron ores (originally found in the country of *Magnesia*, whence the terms "magnet" and "magnetic ore"). By enclosing a bar of soft iron within an insulated wire wound in a helix, a powerful "electro-magnet" is produced when a current is made to pass through the helix; the magnetic

power ceases, however, on shutting off the current, soft iron differing from steel in that the latter can retain magnetic polarity permanently, whilst the former cannot, a fact expressed by saying that steel possesses "coercive force." When brought into proximity of a sufficiently powerful magnet (preferably an electro-magnet) most metals exhibit signs of either attraction or repulsion; those that are attracted by either pole are said to be *paramagnetic;* such as iron (*par excellence*); in a less degree nickel and cobalt; and to a slight extent manganese, chromium, titanium, and palladium, the paramagnetism of some of these being possibly due to the presence of minute traces of iron. Most other metals, and in particular, bismuth, are repelled by both poles, and are said to be *diamagnetic.* When a conducting circuit is made to alter its position in space with reference to a magnet (whether a natural loadstone, a magnetized steel bar, an electro-magnet, or a wire conveying an electric current) an electric current is generated in the circuit, the duration depending on the nature of the relative motions of magnet and circuit.

98. The chief practical applications of these facts are the use of magnetized steel bars for ships' compasses, and the construction of machines whereby powerful electric currents are generated by the motion of conducting wires (driven by steam power) in the vicinity of magnets; the currents thus produced can be applied to electro-metallurgy, and especially to artificial illumination, being far more cheaply obtained than voltaic currents of equal power. Lighthouses have for many years been illuminated in this way; whilst quite recently M. Jablochoff has introduced important improvements in the means whereby the currents are made to produce light, the effect of which is to render this mode of

illumination much more practically useful than before; so much so, that there is a good prospect of electric illumination being hereafter extensively employed for public buildings and many other purposes. One of the most recent inventions depending on these principles is Bell's telephone, a section of which is given in Fig. 33. A permanent steel magnet, A B, is enclosed in a wooden case, a coil of fine insulated wire, C, having

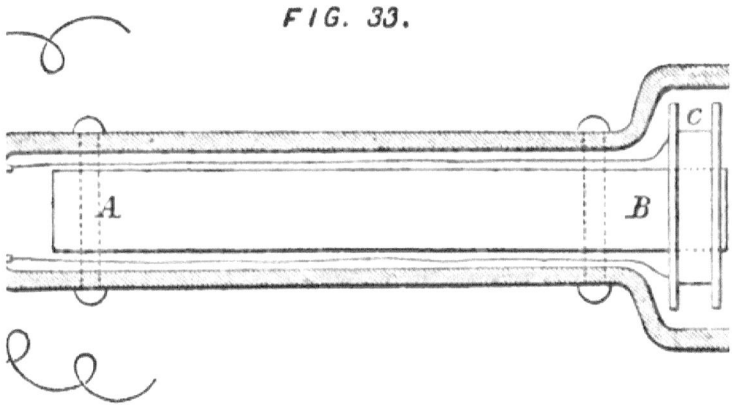

FIG. 33.

been wound round one end of it. A thin plate of iron, D, is fitted in the wooden case just opposite one of the poles of the magnet; when the vibrations of the air caused by speaking in at the mouth-piece, E, strike upon this plate they cause it to vibrate, and hence, as its component particles are in motion in the proximity of the coil of wire, they cause currents to be generated therein; for the soft iron plate is virtually a magnet, being excited by the inductive action of the permanent magnet A B. The two ends of the coil of wire communicate with the binding screws, F G, one of which is "put to earth," (made to communicate with the solid ground by a wire,) the other being connected to the telegraph wire. At the far end (receiving station) is a precisely similar arrangement, one end of the coil of wire being connected to the telegraph wire, and the other put to earth. Whatever currents are generated in the transmitting instrument pass through the telegraph wire to the other instrument, and set up in the vibrating iron plate thereof precisely the same motions as were produced in the vibrating plate of the transmitting instrument, so that the recipient of the message *hears the sounds reproduced* by simply placing his ear to the trumpet mouth of the receiving instrument.

CHAPTER VII.

CHEMICAL RELATIONS OF METALS.

99. **Special applications of Metals.**—Certain metals possess peculiar chemical properties which render them applicable for special purposes for which no other bodies could be so well employed : thus the inactivity of platinum towards acids, oxidising agents, and many other kinds of chemical products, renders it a peculiarly suitable material for the vessels used in many manufacturing operations, such as the concentration of sulphuric acid, the refining of gold in the quartation stage (§ 17), &c. Again, the power which this metal possesses when in a fine state of division to dissolve gases and vapours and cause them to react more energetically on one another leads to its use in the manufacture of certain articles ; thus finely-divided platinum causes the oxidation of sulphur dioxide to sulphuric anhydride when a mixture of sulphur dioxide and oxygen is passed over it ; and a cheap way of producing " Nordhausen sulphuric acid " used by indigo dyers, is based on this principle. The smokeless perfume vaporisers of the shops also depend on this power, the vapour of perfumed alcohol from a wick, when mixed with air, sufficing to keep a little ball of spongy platinum red-hot by the heat developed by the

partial combustion of the vapour in the pores of the platinum, whilst the rest of the spirit is volatilised into the apartment. Döbereiner's lamp, a substitute for matches, in which a current of hydrogen from an automatic generator is turned on to spongy platinum thereby heating the platinum and lighting the jet of hydrogen, depends on the same principle.

100. Again, the vivid combustion of magnesium, due to its readily oxidising, leads to the use of this metal as an illuminant (§ 54); and the solvent power of mercury for precious metals is the foundation of the amalgamation process for their extraction (§§ 16 and 22); whilst the different oxidisability of various metals in air at the ordinary temperature causes the employment of several of them in which this property is least strongly marked to be used as protective coatings for the others that rust more easily but which, on the other hand, are cheaper or more easily worked into the required forms. Thus sheet-iron is coated with tin, or "tinned," forming the tin-plate of the tinman (*fer-blanc*), from which numerous articles of domestic and general use are formed. Similarly iron is coated with zinc, forming "galvanised iron." In practice, both tinning and zincing are carried out by dipping the iron articles (well cleansed and scoured) into cauldrons containing the fused coating metal; the term "galvanised" being due to the former use of voltaic currents for depositing the coating of zinc. By the aid of electricity ornamental and protective coatings of various metals, and in some cases alloys, are deposited on foundations of other material; thus are carried out the electro-gilding and silvering of plate, &c.; the electro-coppering of various metals to give a coppery or bronze-like appearance; the electro-brassing of stereotype plates to enable them to resist wear and tear; the analogous

electro-steeling of copper-plates for engraving; the electro-nickelizing of steel knives, sword blades, &c.; and many similar processes in the arts. By modifying the nature of the surface on which the metal is deposited, it becomes possible to detach therefrom the coating of metal electrically precipitated: on this depends the whole art of *electrotyping;* a cast or mould being prepared of the object to be copied (the surface of the cast being rubbed over with blacklead, or otherwise made to conduct electricity, if this material of the mould be non-conducting), the cast is immersed in the liquid of the depositing cell and connected with the *negative* pole of the electromotor employed, when the metal dissolved in the liquid is deposited.

101. The electro-deposition of gold has almost superseded the old process of so-called *water-gilding*, in which an amalgam of gold is applied to the object to be gilt; by heating, the mercury is volatilized and a dead brownish yellow surface of gold left; by careful burnishing this finally becomes smooth and brilliant. Similarly electro-silvered articles are now used almost to the exclusion of the old forms of plated goods made from bars of copper covered over with thin plates of silver soldered to them, and then rolled out into sheets, &c. which were afterwards worked into the required shapes. It may be noticed in passing that a large portion of the success of the electro-plated articles as compared with the plated copper "Sheffield" goods is due to the use instead of copper of a white or nearly white alloy as ground-work containing copper, nickel, and zinc, (nickel silver). When the silver coating is worn off in places from articles of the old kind, the fact is immediately visible from the red copper showing through; whereas, with the nickel silver goods, the white metal

underneath prevents the loss of the silver coating from being so noticeable. Steel articles are occasionally silvered by a process consisting of the deposition of silver and then heating the mass so as to "burn in" the deposited metal and cause it to adhere more firmly, a fresh coating being deposited on the burnt-in surface, and so on (*Neal's Pyro-silver*) : this encausticating process is equally necessary in the gilding of steel if strong adherence is required.

102. **Alloys.**—The substances produced by the intimate intermixture of metals are sometimes of the nature of distinct chemical compounds, their formation being accompanied by considerable heat-development and their composition being expressible by definite formulæ. In most cases, however, they partake more of the nature of simple mixtures, or of homogeneous solutions converted into the solid state without the separation of particles, differing from one another in physical and chemical nature, of sufficient magnitude to be discernible by a microscope. Nevertheless the formation of these bodies in almost every instance is fairly included in the term "chemical action," according to the usual definition that "chemical action takes place when the properties of the resultant substances are different from those of the original bodies;" for, save in comparatively few instances, the physical and often the chemical properties of an alloy are different from the arithmetical average of those of its components. Thus, according to Matthiessen,[1] all metals excepting zinc, lead, tin, and cadmium, yield alloys, the physical properties of which are not in the proportions calculable from those of the constituents. The physical properties of specific gravity, specific heat, and expansibility are never greatly different in alloys (no

[1] *Chem. Soc. Journal*, 1867, p. 201.

matter what the constituents) from those calculable from the constitution; but the properties of fusibility, crystalline form, conductivity for heat, electricity, and sound, elasticity, tenacity, &c., are invariably considerably different from those calculable from the composition (save only alloys containing no constituents other than zinc, tin, lead, or cadmium).

103. The following tables give a general idea of the average composition of many of the more important alloys in common use for various purposes.

	Coinage Alloys.				Bronze.	Solders.	
	Standard Gold.		Standard Silver.			For Gold.	For Silver.
	English.	French.	English.	French.			
Gold	91·667	90	—	—	—	60	—
Silver	8·333	10	92·5	90	—	25	67
Copper			7·5	10	95	15	22
Tin	—	—	—	—	4	—	—
Zinc	—	—	—	—	1	—	11
	100·000	100	100·0	100	100	100	100

Copper Alloys.

(I.) Brass and Allied Alloys.

	Tombak.	Casting Brass.	Dutch Metal (or-molu).	Brass for lathe-work.	Fine Brass.	Muntz's Metal.	Spelter Solder.	Aich's Gun-metal and Gedge's metal.
Copper	80—95	65—72	70—85	60—70	65	60	50	60—62
Zinc	5—20	35—28	15—25	28—38	35	39	50	36—38
Lead	—	—	0—3	2	—	1	—	—
Tin	—	—	0—3	—	—	—	—	—
Iron	—	—	—	—	—	—	—	2

(II.) Bronzes and Allied Alloys.

	Gun-metal.	Bell-metal.	Speculum metal.	Antique Bronzes.	Medal Bronze.	Casting Bronze.
Copper	85—92	65—80	60	70—95	93	82—83
Tin	8—15	20—35	30	8—15	7	1—3
Antimony	—	0—2	—	—	—	—
Arsenic	—	—	10	—	—	—
Lead	—	—	—	0—1	—	trace—3
Zinc	—	—	—	0—15	—	17—18

(III.) Copper, Nickel, Zinc Alloys, &c.

	German Silver (yellowish).	Chinese Pakfong.	White Nickel Silver.	German Silver for Casting.	Aluminium Bronze.
Copper	50—60	40	50	49	90—95
Zinc	25—30	25	25	24	—
Nickel	15—20	32	25	24	—
Iron	—	3	—	—	—
Lead	—	—	—	3	—
Aluminium	—	—	—	—	5—10

Lead and Tin Alloys.

	Solders.			Pewter.			Britannia Metal.	Queen's Metal.	Type Metal.	Fusible Metal.	Shot.
	Hard.	Ordinary.	Soft.	Pewterer's.	Common.	Finest.					
Lead	67	50	33	50	20	—	—	8	75—85	25	98
Tin	33	50	67	25	80	90	88—92	76	0—5	25	—
Antimony	—	—	—	—	—	7	8—12	8	12—25	—	—
Bismuth	—	—	—	25	—	2	—	8	—	50	—
Copper	—	—	—	—	—	2	—	—	—	—	—
Arsenic	—	—	—	—	—	—	—	—	—	—	2

In many instances the precise combination of metals employed for any particular purpose is regarded as a valuable trade secret by the manufacturer, so that the product preferred by one is not necessarily identical with that used by another. Thus, the character of the alloys pewter, Britannia metal, and the like, vary with the manufacturer, and the purpose for which the alloy is intended, &c.; the pewter for tankards (hard pewter) and that employed for covering over public-house bars, &c. (counter metal), are generally prepared by adding to the tin which forms the basis different amounts of other metals (lead, antimony, bismuth) constituting what is termed "temper." Similarly very different "tempers" are employed to add to tin to form "Britannia metal," "Queen's metal," &c., by different manufacturers; whilst the exact composition of bell-metal is subject to considerable variations according to the size of the bell, &c.

104. In the practical manufacture of alloys, as a general rule, the materials are melted (either separately or together), well incorporated, and then cast into ingots or such other form as may be required; some few alloys, however, are prepared in other ways; for example, brass may be obtained by heating copper-plates with a mixture of calamine and small coal, which gives off zinc in the form of vapour; this zinc is absorbed by the copper and retained by it much as a sheet of gold heated over mercury absorbs the mercurial vapour and becomes whitened in consequence. Certain alloys are obtained directly from complex ores, e.g., spiegeleisen. The standard alloys used for coinage are prepared by incorporating together the right quantities of the ingredients and casting into bars which are rolled into ribbons of the requisite thickness for the kind of coin required, the degree of thickness being judged by punching out discs or "blanks" and

weighing them. When the sheet has been rolled to the requisite extent blanks are punched out by machinery, and are then (after undergoing softening, annealing, and other processes) stamped in the coining press, and passed through a machine which weighs and sorts the coins into those of the just weight (or rather which fall within a slight difference from the true weight, termed the "remedy"), those which are too light and those which are too heavy. These incorrect pieces are melted up again and the metal employed a second time.

Great care must be taken in the selection of the copper used for alloying with the gold or silver, and also purity of the precious metal employed is essential, as minute quantities of foreign metals in the coinage-alloy impair its softness, and either prevent it from rolling or taking the impression from the dies, or render it brittle.

105. The verification of the correctness of the mixture of metals employed (*assaying*) is a most important part of minting operations. In the case of gold coins the process employed depends on the circumstances that whilst neither gold nor silver will oxidize in hot air, other metals, especially in presence of a certain quantity of lead, will do so, so that a separation of precious and common metals may be thus effected with due precautions; also that from an alloy of gold and silver containing not less than two or three parts of silver to one of gold hot nitric acid will dissolve out the silver, leaving behind the whole of the gold (*Quartation*). The metal to be assayed is accurately weighed, pure silver to the extent of two or three times its weight added, and the whole rolled up in a piece of pure sheet lead weighing from six to ten times the amount of gold present (the quantity of lead varying with the amount of foreign alloy). The whole is then heated in a "cupel" (made of bone-ash

CHIEF INDUSTRIAL APPLICATIONS.

compressed in a mould) placed in a "muffle," by a furnace. The fused mass oxidises on the surface, the lead, whilst oxidising, causing copper and other base metals present to oxidise too; the fused oxides are absorbed by the porous cupel, and finally a button of pure gold and silver is left. This is rolled out into a ribbon, annealed, twisted up into a "cornet," and acted on by hot nitric acid, whereby the silver is dissolved out and the pure gold left. After annealing, the residual gold cornet is weighed, and thus the weight of gold present known. If silver be also present, another portion is cupelled with lead without the addition of pure silver; the difference between the weight of the button thus obtained and the pure gold previously found representing the silver present.

Silver alloys may be similarly cupelled with lead; or may be dissolved in nitric acid, and the silver estimated by precipitation with a soluble chloride, the silver chloride thrown down being collected and weighed, or else the bulk of a chloride solution of known strength required to completely precipitate the silver being noted.

106. Certain of the metals and alloys commonly used for household and other purposes, give rise, by their corrosion, to poisonous substances; and in consequence the contact of articles made of such materials with alimentary products not unfrequently gives rise to illness, and sometimes produces fatal results. Lead, copper, and zinc are the metals most injurious in this respect, especially the first. When most kinds of drinking-water are allowed to stand in leaden pipes or cisterns a certain amount of the noxious metal is taken into solution, or becomes suspended in the water; wherefore, whenever water is drawn from a housepipe for drinking purposes the portion which first runs out should be rejected as being liable to contain lead from the pipe; whilst no

water from lead cisterns should ever be used for culinary or potable purposes. Tinned meats, fruits, and the like, are apt to contain lead from the presence of that metal in the impure tin used in making the tinplate canisters, or from that in the soldering junctions. Beer, &c., that has stood in the tubes of a beer-engine all night is apt to contain lead, and cases of lead-poisoning are on record from the continued use of the ale first drawn in the morning. Similarly soda-water, and aërated waters generally, as well as artificially-prepared sparkling wines (more commonly met with than is supposed), are apt to contain lead from the aërating apparatus, whilst pickles, &c., sometimes contain it from the pewter capsule of the bottles in which they are sold. Brass and copper stewpans occasionally become verdigrised, whilst preserved peas, pickles, and the like, are often purposely treated with copper compounds to improve the colour. For the same reason many cooks habitually boil greens for the table with a rusty pennypiece or brass or copper bolt, &c. Acid or saccharine liquids sometimes act on galvanized iron vessels, taking up zinc therefrom. In order to diminish some of these risks, tubes are now manufactured for water-supply, beer-engines, and the like, in which the outside portion of the pipe is composed of lead, but the internal lining is of pure block tin, so that liquids passing through such pipes cannot take up lead unless the tin lining be first eaten through, which is unlikely in most cases, from the less tendency to oxidise and rust exhibited by tin. Even tin, however, is not always free from objection on the score of corrosion under certain circumstances, and consequent impregnation of food with deleterious metallic compounds; the author has recently met with instances of this action in the case of preserved fruits &c., put up in the ordinary tinned canisters ; an unpleasant metallic taste was noticeable, and on analysis considerable amounts of *tin* were detected in solution in the juice, besides iron from the underlying metal of the tinplate : in one case violent colic and diarrhœa lasting for several hours was produced in the persons partaking of the preserved fruits: this corrosive action on tin has only been noticed as yet in the case of very *acid* fruits, &c., such as apples and rhubarb.

107. **Compounds of Metals with Non-metallic Elements.**—Independently of the valuable services rendered by metals in the reguline condition, and by their compounds with one another (alloys), a vast multitude of substances are used for various industrial and general purposes which are of a compound character, the constituent elements being partly metallic, partly non-metallic in their nature. A great number of these bodies are of the nature of what are termed "salts" and are used in numerous manufactures, *e.g.*, potassium and sodium salts in glass-making, soap-making, dyeing, &c. and for medicinal and general purposes. Calcium carbonate in the forms of marble, limestone, chalk, coral, &c., plays an important part in the building and constructive arts, either as a masonry material, or as furnishing lime for mortar and cements. Various natural aluminium silicates in a greater or less degree of purity and in a more or less altered state from the decomposing action of the weather are the basis of the different forms of clay so extensively used for brick-making, pottery, drainage tubes, roofing tiles, terra-cotta work, porcelain, and the like purposes ; whilst mixtures of silicates of certain other metals when artificially prepared by fluxing together sand, lime, soda-ash, &c., with certain variable constituents mainly of a metallic character, constitute the various kinds of glasses, the *colours* of

which (due to the power of absorbing certain kinds of light to a greater extent than others) depend on the character of the metallic silicates formed; enamels and glazes are analogous substances formed on the surface of metals, porcelain, &c. The photographic art would be comparatively of little practical value without silver nitrate, and other silver compounds thence derived; and hundreds of other processes and trades depend more or less on the use of saline metallic compounds of one kind or another.

108. One of the most widespread uses of metallic derivatives is their employment as *pigments* or paints, owing to their power of reflecting certain kinds of light better than others, so that certain tints predominate in the reflected light. Many of these are prepared by "reactions of double decomposition" (§ 5), two soluble compounds when mixed together in the state of solution reacting so as to form an insoluble powder (*precipitate*), and a second salt which remains dissolved and is separated by settling, decantation of supernatant fluid, and washing. Thus if sulphuretted hydrogen and cadmium chloride solutions be mixed, the result is the formation of insoluble yellow cadmium sulphide (the "cadmium yellow" of the painter) and hydrochloric acid which remains in solution; thus :—

$$CdCl_2 + H_2S = CdS + 2HCl.$$

In this way such substances as the various kinds of Prussian and Chinese blues, chromes, arsenical greens, &c., are prepared. Other pigments are formed by dry processes; thus vermilion is produced by heating together mercury and sulphur; ultramarine by heating a mixture of China clay, sodium carbonate, and sulphur; smalt by stamping or grinding to impalpable powder an intensely blue impure glass prepared by fusing cobaltaceous minerals and siliceous matters; and so on with many others.

109. The substance termed "whitelead" is perhaps the most important of bodies of this class, as owing to the peculiar nature of the action of this substance on linseed and other oils it is generally employed as the basis of ordinary "paint," the colour being given by the further incorporation of some suitable pigment. This body is essentially a carbonate or basic carbonate of lead, and may be prepared in several ways: by adding to a soluble salt of lead a soluble carbonate, lead carbonate is thrown down as a heavy white powder. As thus prepared, however, whitelead is crystalline; and this circumstance prevents its possessing as much "body" or covering power, as whitelead made in other ways, e.g., by the "Dutch process," which consists of exposing lead in sheets or gratings, placed in pots with a little acetic acid at the bottom, to the action of the gases given off from spent tan or other fermentable vegetable refuse (stable manure is apt to discolour the product from its evolving sulphuretted hydrogen). In the warm moist atmosphere charged with acetic acid vapour and carbon dioxide produced by heaping up in successive layers a quantity of the pots covered over with boards, and the fermenting tan, the lead rapidly becomes converted into a white mass of basic carbonate which is well washed and separated by elutriation from particles of metallic lead, &c. Recently a process for preparing whitelead and other products directly from galena or complex ores containing lead sulphide has been patented by Fitzgerald and Molloy; this consists of treating the crushed and partially roasted ore with sulphuric acid containing nitric acid; the nitrous fumes evolved being absorbed by sulphuric

acid with access of air whereby a nitric-sulphuric acid is formed which is used over again for a new batch. Any metals present which form soluble sulphates (*e.g.*, copper, iron, zinc) are thus obtained in solution whilst lead sulphate and gangue remain as an insoluble mass (should silver be present, this is retained as insoluble chloride by addition of common salt). The liquid solution of sulphates with excess of acid is boiled down to a smaller bulk, whereby anhydrous metallic sulphates separate, the supernatant acid being used to absorb nitrous fumes and so being employed over again. From the anhydrous sulphates thus obtained the copper is isolated by simply dissolving in water and adding scrap-iron. The insoluble mixture of lead sulphate, silver chloride, and gangue is treated with hot brine to dissolve out silver chloride, and the residue is boiled with salt and a little sulphuric acid whereby sulphate of soda and lead chloride are formed : from the recrystallised lead chloride whitelead is prepared by simply boiling with chalk : this whitelead is stated to be quite free from crystals and to possess admirable body.

A substance much akin to whitelead is *Pattinson's oxychloride of lead* prepared from galena by conversion into chloride by the action of hydrochloric acid, and addition to the warm solution of the well purified chloride of limewater, whereby, with due precautions, a white pigment possessing great body is precipitated.

110. Sulphate of Barium is largely used as an adulterant of whitelead ; for some purposes, however, its addition is advantageous rather than the reverse, as it is not affected by sulphur compounds, which render whitelead black and discoloured ; to avoid this inconvenience, *zinc white* (oxide of zinc) is often used instead of whitelead for painting rooms when the walls, &c., are apt to

be exposed to sewer gases or other sources of sulphuretted hydrogen. Tungstate of zinc and tungstate of barium have been of late years employed for the same reason.[1] Both zinc white and tungsten white are inferior to good whitelead in "body," and do not always blend with oil so as to make as satisfactory a liquid paint as that formed by whitelead; so that, except for painting premises, &c., liable to be discoloured, they are not as largely used as their comparatively non-poisonous properties would render desirable, the processes of whitelead making, and painter's work generally, being peculiarly liable to produce saturnine disease in the workmen employed (to a large extent owing to want of due care on their part.)

111. Closely allied to pigments are the substances termed *Lakes*, many of which indeed are in use as pigments. When colouring matters of animal and vegetable origin, as well as certain artificial dyestuffs from coal-tar products, &c., are dissolved in water, and a metallic salt added, and finally some body which will precipitate the metallic hydrate or other compound, the organic colouring matter is often retained in combination by the precipittated metallic compound, so that a precipitate of complex character is obtained often possessing some marked and brilliant colour. It is noteworthy that many metallic oxides entirely change the colour of the original dye to something quite different. Aluminium lakes are most frequently employed; thus carmine is an alumina lake

---

[1] Various shades of green, yellow, and blue pigments, as well as bronzes, are obtainable from tungsten. Recently, improvements have been made in the manufacture of these colours by Dr. Versmann; and, as considerable quantities of tungsten-containing minerals are raised in certain mines—for the most part not utilised—there is a prospect of tungsten colours being hereafter largely used.

prepared from cochineal; if stannic chloride be also present a scarlet lake is formed.

Certain metallic oxides, especially those of iron and aluminium and tin, possess the power of chemically uniting with organic fibres and also of simultaneously forming lakes; the effect of placing a calico or cloth impregnated with a metallic compound of such a character in a solution of a dyestuff consequently is that the dyestuff is permanently united with the lake by the "mordanting" action of the metallic oxide; although without the oxide the colour would be readily washed out of the fabric, yet when mordanted on, it firmly resists the detergent power of washing. By printing on calicoes patterns in mordant liquors thickened with gum, starch, &c., and then immersing in a dye-vat, numerous colours may be developed at once on the same cloth by one and the same dyestuff, the difference being due to the application to different parts of the fabric of mordants of different compositions. Thus iron liquor gives shades varying from black and dark purple to lilac according to the strength; acetate of alumina similarly produces various reds; a mixture of iron and alumina gives a claret colour. The large trade of calico-printing owes its existence entirely to this circumstance; similarly the scarlet army cloth is prepared by impregnating the cloth with a tin liquor and then boiling with cochineal, when the scarlet tin-cochineal colour is mordanted on to the cloth.

To this brief enumeration of the more important uses of metals and their derivatives hundreds of other applications might be added did space permit; so few substances indeed are there in common use in every day life which are not either made in whole or in part of metal or alloy or other metallic compounds, or produced by the indispensable aid of metallic tools or appliances, that the period when man first began to emerge from the barbarism of the primeval anthropoid may fairly be dated as the time when the utilisation of the natural sources of metals, and the extraction thence of these bodies or of their compounds, first began to be practised.

THE END.

January 1878.

A CATALOGUE

OF

EDUCATIONAL BOOKS,

PUBLISHED BY

MACMILLAN AND CO.,

BEDFORD STREET, STRAND, LONDON.

CLASSICAL.

ÆSCHYLUS—*THE EUMENIDES.* The Greek Text, with Introduction, English Notes, and Verse Translation. By BERNARD DRAKE, M.A., late Fellow of King's College, Cambridge. 8vo. 3s. 6d.

ARISTOTLE—*AN INTRODUCTION TO ARISTOTLE'S RHETORIC.* With Analysis, Notes and Appendices. By E. M. COPE, Fellow and Tutor of Trinity College, Cambridge. 8vo. 14s.

ARISTOTLE ON FALLACIES; OR, THE SOPHISTICI ELENCHI. With Translation and Notes by E. POSTE, M.A. Fellow of Oriel College, Oxford. 8vo. 8s. 6d.

ARISTOPHANES—*THE BIRDS.* Translated into English Verse, with Introduction, Notes, and Appendices, by B. H. KENNEDY, D.D., Regius Professor of Greek in the University of Cambridge. Crown 8vo. 6s.

BELCHER—*SHORT EXERCISES IN LATIN PROSE COMPOSITION AND EXAMINATION PAPERS IN LATIN GRAMMAR*, to which is prefixed a Chapter on Analysis of Sentences. By the Rev. H. BELCHER, M.A., Assistant Master in King's College School, London. Third Edition. 18mo. 1s. 6d. Key, 1s. 6d.

BLACKIE—*GREEK AND ENGLISH DIALOGUES FOR USE IN SCHOOLS AND COLLEGES.* By JOHN STUART BLACKIE, Professor of Greek in the University of Edinburgh. Second Edition. Fcap. 8vo. 2s. 6d.

a

CICERO—*THE SECOND PHILIPPIC ORATION.* With Introduction and Notes. From the German of KARL HALM. Edited, with Corrections and Additions, by Professor JOHN E. B. MAYOR, M.A. Fellow and Classical Lecturer of St. John's College, Cambridge. Fourth Edition, revised. Fcap. 8vo. 5s.

THE ORATIONS OF CICERO AGAINST CATILINA. With Notes and an Introduction from the German of KARL HALM, with additions by Professor A. S. WILKINS, M.A., Owens College, Manchester. Fourth Edition. Fcap. 8vo. 3s. 6d.

THE ACADEMICA OF CICERO. The Text revised and explained by JAMES REID, M.A., Assistant Tutor and late Fellow of Christ's College, Cambridge. Fcap. 8vo. 4s. 6d.

DEMOSTHENES—*ON THE CROWN*, to which is prefixed *ÆSCHINES AGAINST CTESIPHON.* The Greek Text with English Notes. By B. DRAKE, M.A., late Fellow of King's College, Cambridge. Fifth Edition. Fcap. 8vo. 5s.

ELLIS—*PRACTICAL HINTS ON THE QUANTITATIVE PRONUNCIATION OF LATIN*, for the use of Classical Teachers and Linguists. By A. J. ELLIS, B.A., F.R.S. Extra fcap. 8vo. 4s. 6d.

GOODWIN—*SYNTAX OF THE MOODS AND TENSES OF THE GREEK VERB.* By W. W. GOODWIN, Ph.D. New Edition, revised. Crown 8vo. 6s. 6d.

"This scholarly and exhaustive work."—SCHOOL BOARD CHRONICLE.

GREENWOOD—*THE ELEMENTS OF GREEK GRAMMAR*, including Accidence, Irregular Verbs, and Principles of Derivation and Composition; adapted to the System of Crude Forms. By J. G. GREENWOOD, Principal of Owens College, Manchester. Fifth Edition. Crown 8vo. 5s. 6d.

HODGSON—*MYTHOLOGY FOR LATIN VERSIFICATION.* A brief Sketch of the Fables of the Ancients, prepared to be rendered into Latin Verse for Schools. By F. HODGSON, B.D., late Provost of Eton. Fourth Edition, revised by F. C. HODGSON, M.A. 18mo. 3s.

HOMERIC DICTIONARY. For Use in Schools and Colleges. Translated from the German of Dr. G. Autenrieth, with Additions and Corrections by R. P. KEEP, Ph.D. With numerous Illustrations. Crown 8vo. 6s.

HOMER'S ODYSSEY—*THE NARRATIVE OF ODYSSEUS.* With a Commentary by JOHN E. B. MAYOR, M.A., Kennedy Professor of Latin at Cambridge. Part I. Book IX.—XII. Fcap. 8vo. 3s.

HORACE—*THE WORKS OF HORACE*, rendered into English Prose, with Introductions, Running Analysis, and Notes, by J. LONSDALE, M.A., and S. LEE, M.A. Globe 8vo. 3s. 6d.

THE ODES OF HORACE IN A METRICAL PARAPHRASE. By R. M. HOVENDEN. Extra fcap. 8vo. 4s.

HORACE'S LIFE AND CHARACTER. An Epitome of his Satires and Epistles. By R. M. HOVENDEN. Extra fcap. 8vo. 4s. 6d.

JACKSON—*FIRST STEPS TO GREEK PROSE COMPOSITION.* By BLOMFIELD JACKSON, M.A. Assistant-Master in King's College School, London. Fourth Edition, revised and enlarged. 18mo. 1s. 6d.

"A capital little book for beginners."—SPECTATOR.

JEBB—Works by R. C. JEBB, M.A., Professor of Greek in the University of Glasgow.

THE ATTIC ORATORS FROM ANTIPHON TO ISAEOS. 2 vols. 8vo. 25s.

THE CHARACTERS OF THEOPHRASTUS. Translated from a Revised Text, with Introduction and Notes. Extra fcap. 8vo. 6s. 6d.

JUVENAL—*THIRTEEN SATIRES OF JUVENAL.* With a Commentary. By JOHN E. B. MAYOR, M.A., Kennedy Professor of Latin at Cambridge. Second Edition, enlarged. Vol. I. Crown 8vo. 7s. 6d. Or Parts I. and II. 3s. 6d. each.

LYSIAS—*SELECT ORATIONS.* Edited, with Notes, &c., by E. S. SHUCKBURGH. [*In preparation.*

MARSHALL — *A TABLE OF IRREGULAR GREEK VERBS*, classified according to the arrangement of Curtius' Greek Grammar. By J. M. MARSHALL, M.A., one of the Masters in Clifton College. 8vo. cloth. Third Edition. 1s.

MAYOR (JOHN E. B.) — *FIRST GREEK READER*. Edited after KARL HALM, with Corrections and large Additions by Professor JOHN E. B. MAYOR, M.A., Fellow and Classical Lecturer of St. John's College, Cambridge. Third Edition, revised. Fcap. 8vo. 4s. 6d.

BIBLIOGRAPHICAL CLUE TO LATIN LITERATURE. Edited after HÜBNER, with large Additions by Professor JOHN E. B. MAYOR. Crown 8vo. 6s. 6d.

"An extremely useful volume that should be in the hands of all scholars."—ATHENÆUM.

MAYOR (JOSEPH B.) — *GREEK FOR BEGINNERS*. By the Rev. J. B. MAYOR, M.A., Professor of Classical Literature in King's College, London. Part I., with Vocabulary, 1s. 6d. Parts II. and III., with Vocabulary and Index, 3s. 6d. complete in one Vol. New Edition. Fcap. 8vo. cloth. 4s. 6d.

NIXON — *PARALLEL EXTRACTS* arranged for translation into English and Latin, with Notes on Idioms. By J. E. NIXON, M.A., Classical Lecturer, King's College, London. Part I.—Historical and Epistolary. Second Edition, revised and enlarged. Crown 8vo. 3s. 6d.

A FEW NOTES ON LATIN RHETORIC. With Tables and Illustrations. By J. E. NIXON, M.A. Crown 8vo. 2s.

PEILE (JOHN, M.A.) — *AN INTRODUCTION TO GREEK AND LATIN ETYMOLOGY.* By JOHN PEILE, M.A., Fellow and Tutor of Christ's College, Cambridge, formerly Teacher of Sanskrit in the University of Cambridge. Third and Revised Edition. Crown 8vo. 10s. 6d.

"A very valuable contribution to the science of language."—SATURDAY REVIEW.

CLASSICAL. 5

PLATO—*THE REPUBLIC OF PLATO.* Translated into English, with an Analysis and Notes, by J. LL. DAVIES, M.A., and D. J. VAUGHAN, M.A. Third Edition, with **Vignette** Portraits of Plato and Socrates, engraved by JEENS from an Antique Gem. 18mo. 4s. 6d.

PLAUTUS—*THE MOSTELLARIA OF PLAUTUS.* With Notes, Prolegomena, and Excursus. By WILLIAM RAMSAY, M.A., formerly Professor of Humanity in the University of Glasgow. Edited by Professor GEORGE G. RAMSAY, **M.A.**, of the University of Glasgow. 8vo. 14s.

POTTS (A. W., M.A.)—*HINTS TOWARDS LATIN PROSE COMPOSITION.* By ALEXANDER W. POTTS, M.A., LL.D., late Fellow of St. John's College, Cambridge; Assistant Master in Rugby School; and Head Master of the Fettes College, Edinburgh. New Edition. Extra fcap. 8vo. 3s.

ROBY—*A GRAMMAR OF THE LATIN LANGUAGE,* from Plautus to Suetonius. By H. J. ROBY, M.A., late Fellow of St. John's College, Cambridge. In Two Parts. Third Edition. Part I. containing:—Book I. Sounds. Book II. Inflexions. Book III. Word-formation. Appendices. Crown 8vo. 8s. 6d. Part II.—Syntax, Prepositions, &c. Crown 8vo. 10s. 6d.

"Marked by the clear and practised insight of a master in his **art.** A book that would do honour to any country."—ATHENÆUM.

RUST—*FIRST STEPS TO LATIN PROSE COMPOSITION.* By the Rev. G. RUST, M.A. of Pembroke College, Oxford, Master of the Lower School, King's College, London. Fifth Edition. 18mo. 1s. 6d.

SALLUST—*CAII SALLUSTII CRISPI CATILINA ET JUGURTHA.* For use in Schools. With copious Notes. By C. MERIVALE, B.D. New Edition, carefully revised and enlarged. Fcap. 8vo. 4s. 6d. Or separately, 2s. 6d. each.

TACITUS—*THE HISTORY OF TACITUS TRANSLATED INTO ENGLISH.* By A. J. Church, M.A., and W. J. Brodribb, M.A. With Notes and a Map. Third Edition. Crown 8vo. 6s.

"A scholarly and faithful translation."—Spectator.

THE AGRICOLA AND GERMANIA OF TACITUS. A Revised Text, English Notes, and Maps. By A. J. Church, M.A., and W. J. Brodribb, M.A. New Edition. Fcap. 8vo. 3s. 6d. Or separately, 2s. each.

"A model of careful editing, being at once compact, complete, and correct, as well as neatly printed and elegant in style."—Athenæum.

THE AGRICOLA AND GERMANY, WITH THE DIALOGUE ON ORATORY. Translated into English by A. J. Church, M.A., and W. J. Brodribb, M.A. With Maps and Notes. New and Revised Edition. Crown 8vo. 4s. 6d.

THE ANNALS. Translated, with Notes and Maps, by A. J. Church and W. J. Brodribb. Second Edition. Crown 8vo. 7s. 6d.

THE ANNALS. Book VI. By the same Editors. With Notes. [*Nearly ready.*

TERENCE—*HAUTON TIMORUMENOS.* Edited, with Introduction and Notes, by E. S. Shuckburgh, M.A. Fcap. 8vo. 3s. With Translation, 3s. 6d.

THEOPHRASTUS—*THE CHARACTERS OF THEOPHRASTUS.* An English Translation from a Revised Text. With Introduction and Notes. By R. C. Jebb, M.A., Professor of Greek in the University of Glasgow. Extra fcap. 8vo. 6s. 6d.

"A very handy and scholarly edition."—Saturday Review.

THRING—Works by the Rev. E. THRING, M.A., Head Master of Uppingham School.

A LATIN GRADUAL. A First Latin Construing Book for Beginners. New Edition, enlarged, with Coloured Sentence Maps. Fcap. 8vo. 2s. 6d.

A MANUAL OF MOOD CONSTRUCTIONS. Fcap. 8vo. 1s. 6d.

A CONSTRUING BOOK. Fcap 8vo. 2s. 6d.

CLASSICAL.

THUCYDIDES—*BOOKS VI. AND VII.*, with Notes. Fifth Edition, revised and enlarged, with Map. By the Rev. PERCIVAL FROST, M.A. Fcap. 8vo. 5s.

VIRGIL—*THE WORKS OF VIRGIL RENDERED INTO ENGLISH PROSE*, with Notes, Introductions, Running Analysis, and an Index, by JAMES LONSDALE, M.A., and SAMUEL LEE, M.A. Second Edition. Globe 8vo. 3s. 6d.; gilt edges, 4s. 6d.

"A more complete edition of Virgil in English it is scarcely possible to conceive than the scholarly work before us."—GLOBE.

WRIGHT—Works by J. WRIGHT, M.A., late Head Master of Sutton Coldfield School.

HELLENICA; OR, A HISTORY OF GREECE IN GREEK, as related by Diodorus and Thucydides; being a First Greek Reading Book, with explanatory Notes, Critical and Historical. Third Edition with a Vocabulary. Fcap. 8vo. 3s. 6d.

A HELP TO LATIN GRAMMAR; or, The Form and Use of Words in Latin, with Progressive Exercises. Crown 8vo. 4s. 6d.

THE SEVEN KINGS OF ROME. An Easy Narrative, abridged from the First Book of Livy by the omission of Difficult Passages; being a First Latin Reading Book, with Grammatical Notes. Fifth Edition. With Vocabulary, 3s. 6d.

FIRST LATIN STEPS; OR, AN INTRODUCTION BY A SERIES OF EXAMPLES TO THE STUDY OF THE LATIN LANGUAGE. Crown 8vo. 5s.

ATTIC PRIMER. Arranged for the Use of Beginners. Extra fcap. 8vo. 4s. 6d.

A COMPLETE LATIN COURSE, comprising Rules with Examples, Exercises, both Latin and English, on each Rule, and Vocabularies. Crown 8vo. 4s. 6d.

MATHEMATICS.

AIRY—Works by Sir G. B. AIRY, K.C.B., Astronomer Royal :—

ELEMENTARY TREATISE ON PARTIAL DIFFERENTIAL EQUATIONS. Designed for the Use of Students in the Universities. With Diagrams. Second Edition. Crown 8vo. 5s. 6d.

ON THE ALGEBRAICAL AND NUMERICAL THEORY OF ERRORS OF OBSERVATIONS AND THE COMBINATION OF OBSERVATIONS. Second Edition, revised. Crown 8vo. 6s. 6d.

UNDULATORY THEORY OF OPTICS. Designed for the Use of Students in the University. New Edition. Crown 8vo. 6s. 6d.

ON SOUND AND ATMOSPHERIC VIBRATIONS. With the Mathematical Elements of Music. Designed for the Use of Students in the University. Second Edition, Revised and Enlarged. Crown 8vo. 9s.

A TREATISE OF MAGNETISM. Designed for the Use of Students in the University. Crown 8vo. 9s. 6d.

AIRY (OSMUND)—*A TREATISE ON GEOMETRICAL OPTICS.* Adapted for the use of the Higher Classes in Schools. By OSMUND AIRY, B.A., one of the Mathematical Masters in Wellington College. Extra fcap. 8vo. 3s. 6d.

BAYMA—*THE ELEMENTS OF MOLECULAR MECHANICS.* By JOSEPH BAYMA, S.J., Professor of Philosophy, Stonyhurst College. Demy 8vo. 10s. 6d.

MATHEMATICS. 9

BEASLEY—*AN ELEMENTARY TREATISE ON PLANE TRIGONOMETRY.* With Examples. By R. D. BEASLEY, M.A., Head Master of Grantham Grammar School. Fifth Edition, revised and enlarged. Crown 8vo. 3s. 6d.

BLACKBURN (HUGH)—*ELEMENTS OF PLANE TRIGONOMETRY*, for the use of the Junior Class in Mathematics in the University of Glasgow. By HUGH BLACKBURN, M.A., Professor of Mathematics in the University of Glasgow. Globe 8vo. 1s. 6d.

BOOLE—Works by G. BOOLE, D.C.L., F.R.S., late Professor of Mathematics in the Queen's University, Ireland.

A TREATISE ON DIFFERENTIAL EQUATIONS. Third and Revised Edition. Edited by I. TODHUNTER. Crown 8vo. 14s.

A TREATISE ON DIFFERENTIAL EQUATIONS. Supplementary Volume. Edited by I. TODHUNTER. Crown 8vo. 8s. 6d.

THE CALCULUS OF FINITE DIFFERENCES. Crown 8vo. 10s. 6d. New Edition, revised by J. F. MOULTON.

BROOK-SMITH (J.) *ARITHMETIC IN THEORY AND PRACTICE.* By J. BROOK-SMITH, M.A., LL.B., St. John's College, Cambridge; Barrister-at-Law; one of the Masters of Cheltenham College. New Edition, revised. Crown 8vo. 4s. 6d.

"A valuable Manual of Arithmetic of the Scientific kind. The best we have seen."—LITERARY CHURCHMAN.

CAMBRIDGE SENATE-HOUSE PROBLEMS and RIDERS WITH SOLUTIONS :—

1875—*PROBLEMS AND RIDERS.* By A. G. GREENHILL, M.A. Crown 8vo. 8s. 6d.

CANDLER—*HELP TO ARITHMETIC.* Designed for the use of Schools. By H. CANDLER, M.A., Mathematical Master of Uppingham School. Extra fcap. 8vo. 2s. 6d.

CHEYNE—Works by C. H. H. CHEYNE, M.A., F.R.A.S.

AN ELEMENTARY TREATISE ON THE PLANETARY THEORY. With a Collection of Problems. Second Edition. Crown 8vo. 6s. 6d.

THE EARTH'S MOTION OF ROTATION. Crown 8vo. 3s. 6d.

CHILDE—*THE SINGULAR PROPERTIES OF THE ELLIPSOID AND ASSOCIATED SURFACES OF THE NTH DEGREE.* By the Rev. G. F. CHILDE, M.A., Author of "Ray Surfaces," "Related Caustics," &c. 8vo. 10s. 6d.

CHRISTIE—*A COLLECTION OF ELEMENTARY TEST-QUESTIONS IN PURE AND MIXED MATHEMATICS;* with Answers and Appendices on Synthetic Division, and on the Solution of Numerical Equations by Horner's Method. By JAMES R. CHRISTIE, F.R.S., Royal Military Academy, Woolwich. Crown 8vo. 8s. 6d.

CUMMING—*AN INTRODUCTION TO THE THEORY OF ELECTRICITY.* By LINNÆUS CUMMING, M.A., one of the Masters of Rugby School. With Illustrations. Crown 8vo. 8s. 6d.

CUTHBERTSON—*EUCLIDIAN GEOMETRY.* By FRANCIS CUTHBERTSON, M.A., LL.D., Head Mathematical Master of the City of London School. Extra fcap. 8vo. 4s. 6d.

DALTON—Works by the Rev. T. DALTON, M.A., Assistant Master of Eton College.

RULES AND EXAMPLES IN ARITHMETIC. New Edition. 18mo. 2s. 6d.

Answers to the Examples are appended.

RULES AND EXAMPLES IN ALGEBRA. Part I. Second Edition. 18mo. 2s. Part II. 18mo. 2s. 6d.

MATHEMATICS.

DAY—*PROPERTIES OF CONIC SECTIONS PROVED GEOMETRICALLY.* Part I., THE ELLIPSE, with Problems. By the Rev. H. G. DAY, M.A., Head Master of Sedburgh Grammar School. Crown 8vo. 3s. 6d.

DODGSON—*AN ELEMENTARY TREATISE ON DETERMINANTS*, with their Application to Simultaneous Linear Equations and Algebraical Geometry. By CHARLES L. DODGSON, M.A. Small 4to. 10s. 6d.

DREW—*GEOMETRICAL TREATISE ON CONIC SECTIONS.* By W. H. DREW, M.A., St. John's College, Cambridge. Fifth Edition, enlarged. Crown 8vo. 5s.

SOLUTIONS TO THE PROBLEMS IN DREW'S CONIC SECTIONS. Crown 8vo. 4s. 6d.

EDGAR (J. H.) and PRITCHARD (G. S.)—*NOTE-BOOK ON PRACTICAL SOLID OR DESCRIPTIVE GEOMETRY.* Containing Problems with help for Solutions. By J. H. EDGAR, M.A., Lecturer on Mechanical Drawing at the Royal School of Mines, and G. S. PRITCHARD. Third Edition, revised and enlarged. Globe 8vo. 3s.

FERRERS—Works by the Rev. N. M. FERRERS, M.A., Fellow and **Tutor** of Gonville and Caius College, Cambridge.

AN ELEMENTARY TREATISE ON TRILINEAR CO-ORDINATES, the Method of Reciprocal Polars, and the Theory of Projectors. Third Edition, revised. Crown 8vo. 6s. 6d.

AN ELEMENTARY TREATISE ON SPHERICAL HARMONICS, AND SUBJECTS CONNECTED WITH THEM. Crown 8vo. 7s. 6d.

FROST—Works by PERCIVAL FROST, M.A., formerly Fellow of St. John's College, Cambridge; Mathematical Lecturer of King's College.

AN ELEMENTARY TREATISE ON CURVE TRACING. By PERCIVAL FROST, M.A. 8vo. 12s.

FROST *Continued*—

THE FIRST THREE SECTIONS OF NEWTON'S PRINCIPIA, With Notes and Illustrations. Also a collection of Problems, principally intended as Examples of Newton's Methods. By PERCIVAL FROST, M.A. Second Edition. 8vo. 10s. 6d.

SOLID GEOMETRY. A New Edition, revised and enlarged of the Treatise by FROST and WOLSTENHOLME. In 2 Vols. Vol. I. 8vo. 16s.

GODFRAY—Works by HUGH GODFRAY, M.A., Mathematical Lecturer at Pembroke College, Cambridge.

A TREATISE ON ASTRONOMY, for the Use of Colleges and Schools. New Edition. 8vo. 12s. 6d.

AN ELEMENTARY TREATISE ON THE LUNAR THEORY, with a Brief Sketch of the Problem up to the time of Newton. Second Edition, revised. Crown 8vo. 5s. 6d.

HEMMING—AN ELEMENTARY TREATISE ON THE DIFFERENTIAL AND INTEGRAL CALCULUS, for the Use of Colleges and Schools. By G. W. HEMMING, M.A., Fellow of St. John's College, Cambridge. Second Edition, with Corrections and Additions. 8vo. 9s.

JACKSON — GEOMETRICAL CONIC SECTIONS. An Elementary Treatise in which the Conic Sections are defined as the Plane Sections of a Cone, and treated by the Method of Projection. By J. STUART JACKSON, M.A., late Fellow of Gonville and Caius College, Cambridge. Crown 8vo. 4s. 6d.

JELLET (JOHN H.)—A TREATISE ON THE THEORY OF FRICTION. By JOHN H. JELLET, B.D., Senior Fellow of Trinity College, Dublin; President of the Royal Irish Academy. 8vo. 8s. 6d.

JONES and CHEYNE—ALGEBRAICAL EXERCISES. Progressively Arranged. By the Rev. C. A. JONES, M.A., and C. H. CHEYNE, M.A., F.R.A.S., Mathematical Masters of Westminster School. New Edition. 18mo. 2s. 6d.

MATHEMATICS.

KELLAND and **TAIT**—*INTRODUCTION TO QUATERNIONS*, with numerous examples. By P. KELLAND, M.A., F.R.S.; and P. G. TAIT, M.A., Professors in the department of Mathematics in the University of Edinburgh. Crown 8vo. 7s. 6d.

KITCHENER—*A GEOMETRICAL NOTE-BOOK*, containing Easy Problems in Geometrical Drawing preparatory to the Study of Geometry. For the use of Schools. By F. E. KITCHENER, M.A., Mathemathical Master at Rugby. Third Edition. 4to. 2s.

MAULT—*NATURAL GEOMETRY*: an Introduction to the Logical Study of Mathematics. For Schools and Technical Classes. With Explanatory Models, based upon the Tachymetrical Works of Ed. Lagout. By A. MAULT. 18mo. 1s. Models to Illustrate the above, in Box, 12s. 6d.

MERRIMAN—*ELEMENTS OF THE METHOD OF LEAST SQUARES*. By MANSFIELD MERRIMAN, Ph.D. Crown 8vo. 7s. 6d.

MILLAR—*ELEMENTS OF DESCRIPTIVE GEOMETRY*. By J. B. MILLAR, B.E. Crown 8vo. [*Nearly ready.*

MORGAN—*A COLLECTION OF PROBLEMS AND EXAMPLES IN MATHEMATICS*. With Answers. By H. A. Morgan, M.A., Sadlerian and Mathematical Lecturer of Jesus College, Cambridge. Crown 8vo. 6s. 6d.

NEWTON'S *PRINCIPIA*. Edited by Prof. Sir W. THOMSON and Professor BLACKBURN. 4to. cloth. 31s. 6d.

"Undoubtedly the finest edition of the text of the 'Principia' which has hitherto appeared." EDUCATIONAL TIMES.

PARKINSON—Works by S. PARKINSON, D.D., F.R.S., Tutor and Prælector of St. John's College, Cambridge.

AN ELEMENTARY TREATISE ON MECHANICS. For the Use of the Junior Classes at the University and the Higher Classes in Schools. With a Collection of Examples. Fifth Edition, revised. Crown 8vo. cloth. 9s. 6d.

A TREATISE ON OPTICS. Third Edition, revised and enlarged. Crown 8vo. cloth. 10s. 6d.

14   MACMILLAN'S EDUCATIONAL CATALOGUE.

PHEAR—*ELEMENTARY HYDROSTATICS.* With Numerous Examples. By J. B. PHEAR, M.A., Fellow and late Assistant Tutor of Clare College, Cambridge. Fourth Edition. Crown 8vo. cloth. 5s. 6d.

PIRIE—*LESSONS ON RIGID DYNAMICS.* By the Rev. G. PIRIE, M.A., Fellow and Tutor of Queen's College, Cambridge. Crown 8vo. 6s.

PUCKLE—*AN ELEMENTARY TREATISE ON CONIC SECTIONS AND ALGEBRAIC GEOMETRY.* With Numerous Examples and Hints for their Solution; especially designed for the Use of Beginners. By G. H. PUCKLE, M.A. Fourth Edition, revised and enlarged. Crown 8vo. 7s. 6d.

RAWLINSON—*ELEMENTARY STATICS*, by the Rev. GEORGE RAWLINSON, M.A. Edited by the Rev. EDWARD STURGES, M.A. Crown 8vo. 4s. 6d.

RAYLEIGH—*THE THEORY OF SOUND.* By LORD RAYLEIGH, M.A., F.R.S., formerly Fellow of Trinity College, Cambridge. In 2 Vols. 8vo. Vol. I. 12s. 6d.

[*Vol. II. in the press.*

REYNOLDS—*MODERN METHODS IN ELEMENTARY GEOMETRY.* By E. M. REYNOLDS, M.A., Mathematical Master in Clifton College. Crown 8vo. 3s. 6d.

ROUTH—Works by EDWARD JOHN ROUTH, M.A., late Fellow and Assistant Tutor of St. Peter's College, Cambridge; Examiner in the University of London.

AN ELEMENTARY TREATISE ON THE DYNAMICS OF THE SYSTEM OF RIGID BODIES. With numerous Examples. Third and enlarged Edition. 8vo. 21s.

STABILITY OF A GIVEN STATE OF MOTION, PARTICULARLY STEADY MOTION. Adams' Prize Essay for 1877. 8vo. 8s. 6d.

SMITH—Works by the Rev. BARNARD SMITH, M.A., Rector of Glaston, Rutland, late Fellow and Senior Bursar of St. Peter's College, Cambridge.

ARITHMETIC AND ALGEBRA, in their Principles and Application; with numerous systematically arranged Examples taken from the Cambridge Examination Papers, with especial

SMITH *Continued*—

reference to the Ordinary Examination for the B.A. Degree. Thirteenth Edition, carefully revised. Crown 8vo. 10s. 6d.

"To all those whose minds are sufficiently developed to comprehend the simplest mathematical reasoning, and who have not yet thoroughly mastered the principles of Arithmetic and Algebra, it is calculated to be of great advantage."—ATHENÆUM.

"Mr. Smith's work is a most useful publication. The rules are stated with great clearness. The examples are well selected, and worked out with just sufficient detail, without being encumbered by too minute explanations; and there prevails throughout it that just proportion of theory and practice which is the crowning excellence of an elementary work."—DEAN PEACOCK.

ARITHMETIC FOR SCHOOLS. New Edition. Crown 8vo. 4s. 6d.

"Admirably adapted for insertion, combining just sufficient theory with a large and well-selected collection of exercises for practice."—JOURNAL OF EDUCATION.

A KEY TO THE ARITHMETIC FOR SCHOOLS. New Edition. Crown 8vo. 8s. 6d.

EXERCISES IN ARITHMETIC. Crown 8vo. limp cloth. 2s. With Answers. 2s. 6d.

Or sold separately, Part I. 1s. ; Part II. 1s. ; Answers, 6d.

SCHOOL CLASS-BOOK OF ARITHMETIC. 18mo. cloth. 3s.

Or sold separately, Parts I. and II. 10d. each ; Part III. 1s.

KEYS TO SCHOOL CLASS-BOOK OF ARITHMETIC. Complete in one volume, 18mo. cloth. 6s. 6d. ; or Parts I., II., and III., 2s. 6d. each.

SHILLING BOOK OF ARITHMETIC FOR NATIONAL AND ELEMENTARY SCHOOLS. 18mo. cloth. Or separately, Part I. 2d. ; Part II. 3d. ; Part III. 7d. Answers, 6d.

THE SAME, with Answers complete. 18mo, cloth. 1s. 6d.

KEY TO SHILLING BOOK OF ARITHMETIC. 18mo. 4s. 6d.

EXAMINATION PAPERS IN ARITHMETIC. 18mo. 1s. 6d. The same, with Answers, 18mo. 1s. 9d. Answers, 3d.

SMITH *Continued*—

KEY TO EXAMINATION PAPERS IN ARITHMETIC. 18mo. 4s. 6d.

THE METRIC SYSTEM OF ARITHMETIC, ITS PRINCIPLES AND APPLICATIONS, with numerous Examples, written expressly for Standard V. in National Schools. Third Edition. 18mo. cloth, sewed. 3d.

A CHART OF THE MÉTRIC SYSTEM, on a Sheet, size 42 in. by 34 in. on Roller, 1s. 6d., mounted and varnished price 3s. 6d. **Third** Edition.

"We do not remember that ever we have seen teaching by a chart more happily carried out."—SCHOOL BOARD CHRONICLE.

Also a Small Chart on a Card, price 1d.

EASY LESSONS IN ARITHMETIC, combining Exercises in Reading, Writing, Spelling, and Dictation. Part I. for Standard **I. in** National Schools. Crown 8vo. 9d.

"We should strongly advise every one to study carefully Mr. Barnard Smith's Lessons in Arithmetic, Writing, and Spelling. A more excellent little work for a first introduction to knowledge cannot well be written. Mr. Smith's larger Text-books on Arithmetic and Algebra are already most favourably known, and he has proved now that the difficulty of writing a text-book which begins *ab ovo* is really surmountable; but we shall be **much** mistaken if this little book has not cost its author more thought and **mental** labour than any of his more elaborate text-books. The plan to combine arithmetical lessons with those in reading and spelling is perfectly novel, and **it is** worked out in accordance with the aims of our National Schools; and we are convinced that its general introduction **in** all elementary schools throughout the country will produce great educational advantages."—WESTMINSTER REVIEW.

EXAMINATION CARDS IN ARITHMETIC. (Dedicated to Lord Sandon.) With Answers and Hints.

Standards I. and II. in box, 1s. 6d. Standards III., IV. and V., in boxes, 1s. 6d. each. Standard VI. in Two Parts, in boxes, 1s. 6d. each.

A and B papers, of nearly the same difficulty, **are** given so as to prevent copying, and the Colours of the A and B papers differ in **each** Standard, and from those of every other Standard, so that a master or mistress can see at a glance whether the children have the proper papers.

MATHEMATICS.

SNOWBALL—*THE ELEMENTS OF PLANE AND SPHERICAL TRIGONOMETRY;* with the Construction and Use of Tables of Logarithms. By J. C. SNOWBALL, M.A. Eleventh Edition. Crown 8vo. 7s. 6d.

SYLLABUS OF PLANE GEOMETRY (corresponding to Euclid, Books I.—VI.). Prepared by the Association for the Improvement of Geometrical Teaching. Third Edition. Crown 8vo. 1s.

TAIT and STEELE—*A TREATISE ON DYNAMICS OF A PARTICLE.* With numerous Examples. By Professor TAIT and MR. STEELE. New Edition, enlarged. Crown 8vo. 10. 6d.

TEBAY—*ELEMENTARY MENSURATION FOR SCHOOLS.* With numerous Examples. By SEPTIMUS TEBAY, B.A., Head Master of Queen Elizabeth's Grammar School, Rivington. Extra fcap. 8vo. 3s. 6d.

TODHUNTER—Works by I. TODHUNTER, M.A., F.R.S., of St. John's College, Cambridge.

> "Mr. Todhunter is chiefly known to students of Mathematics as the author of a series of admirable mathematical text-books, which possess the rare qualities of being clear in style and absolutely free from mistakes, typographical or other."—SATURDAY REVIEW.

THE ELEMENTS OF EUCLID. For the Use of Colleges and Schools. New Edition. 18mo. 3s. 6d.

MENSURATION FOR BEGINNERS. With numerous Examples. New Edition. 18mo. 2s. 6d.

ALGEBRA FOR BEGINNERS. With numerous Examples. New Edition. 18mo. 2s. 6d.

KEY TO ALGEBRA FOR BEGINNERS. Crown 8vo. 6s. 6d.

TRIGONOMETRY FOR BEGINNERS. With numerous Examples. New Edition. 18mo. 2s. 6d.

KEY TO TRIGONOMETRY FOR BEGINNERS. Crown 8vo. 8s. 6d.

TODHUNTER *Continued—*

MECHANICS FOR BEGINNERS. With numerous Examples. New Edition. 18mo. 4s. 6d.

ALGEBRA. For the Use of Colleges and Schools. Seventh Edition. Crown 8vo. 7s. 6d.

KEY TO ALGEBRA FOR THE USE OF COLLEGES AND SCHOOLS. Crown 8vo. 10s. 6d.

AN ELEMENTARY TREATISE ON THE THEORY OF EQUATIONS. Third Edition, revised. Crown 8vo. 7s. 6d.

PLANE TRIGONOMETRY. For Schools and Colleges. Sixth Edition. Crown 8vo. 5s.

KEY TO PLANE TRIGONOMETRY. Crown 8vo. 10s. 6d.

A TREATISE ON SPHERICAL TRIGONOMETRY. Third Edition, enlarged. Crown 8vo. 4s. 6d.

PLANE CO-ORDINATE GEOMETRY, as applied to the Straight Line and the Conic Sections. With numerous Examples. Fifth Edition, revised and enlarged. Crown 8vo. 7s. 6d.

A TREATISE ON THE DIFFERENTIAL CALCULUS. With numerous Examples. Seventh Edition. Crown 8vo. 10s. 6d.

A TREATISE ON THE INTEGRAL CALCULUS AND ITS APPLICATIONS. With numerous Examples. Fourth Edition, revised and enlarged. Crown 8vo. 10s. 6d.

EXAMPLES OF ANALYTICAL GEOMETRY OF THREE DIMENSIONS. Third Edition, revised. Crown 8vo. 4s.

A TREATISE ON ANALYTICAL STATICS. With numerous Examples. Fourth Edition, revised and enlarged. Crown 8vo. 10s. 6d.

TODHUNTER—*Continued*—

A HISTORY OF THE MATHEMATICAL THEORY OF PROBABILITY, from the time of Pascal to that of Laplace. 8vo. 18s.

RESEARCHES IN THE CALCULUS OF VARIATIONS, principally on the Theory of Discontinuous Solutions: an Essay to which the Adams Prize was awarded in the University of Cambridge in 1871. 8vo. 6s.

A HISTORY OF THE MATHEMATICAL THEORIES OF ATTRACTION, AND THE FIGURE OF THE EARTH, from the time of Newton to that of Laplace. 2 vols. 8vo. 24s.

AN ELEMENTARY TREATISE ON LAPLACE'S, LAMÉ'S, AND BESSEL'S FUNCTIONS. Crown 8vo. 10s. 6d.

WILSON (J. M.)—*ELEMENTARY GEOMETRY.* Books I. II. III. Containing the Subjects of Euclid's first Four Books. Following the Syllabus of the Geometrical Association. By J. M. Wilson, M.A., late Fellow of St. John's College Cambridge, and Mathematical Master of Rugby School. New Edition. Extra fcap. 8vo. 3s. 6d.

SOLID GEOMETRY AND CONIC SECTIONS. With Appendices on Transversals and Harmonic Division. For the Use of Schools. By J. M. Wilson, M.A. Third Edition. Extra fcap. 8vo. 3s. 6d.

WILSON (W. P.)—*A TREATISE ON DYNAMICS.* By W. P. Wilson, M.A., Fellow of St. John's College, Cambridge, and Professor of Mathematics in Queen's College, Belfast. 8vo. 9s. 6d.

WOLSTENHOLME—*A BOOK OF MATHEMATICAL PROBLEMS*, on Subjects included in the Cambridge Course. By Joseph Wolstenholme, Fellow of Christ's College, sometime Fellow of St. John's College, and lately Lecturer in Mathematics at Christ's College. Crown 8vo. 8s. 6d.

"Judicious, symmetrical, and well arranged."—GUARDIAN.

SCIENCE.

ELEMENTARY CLASS-BOOKS.

ASTRONOMY, by the Astronomer Royal.
POPULAR ASTRONOMY. With Illustrations. By Sir G. B. AIRY, K.C.B., Astronomer Royal. New Edition. 18mo. 4s. 6d.

Six lectures, intended "to explain to intelligent persons the principles on which the instruments of an Observatory are constructed, and the principles on which the observations made with these instruments are treated for deduction of the distances and weights of the bodies of the Solar System."

ASTRONOMY.
ELEMENTARY LESSONS IN ASTRONOMY. With Coloured Diagram of the Spectra of the Sun, Stars, and Nebulæ, and numerous Illustrations. By J. NORMAN LOCKYER, F.R.S. New Edition. 18mo. 5s. 6d.

"Full, clear, sound, and worthy of attention, not only as a popular exposition, but as a scientific 'Index.'"—ATHENÆUM.

QUESTIONS ON LOCKYER'S ELEMENTARY LESSONS IN ASTRONOMY. For the Use of Schools. By JOHN FORBES-ROBERTSON. 18mo. cloth limp. 1s. 6d.

PHYSIOLOGY.
LESSONS IN ELEMENTARY PHYSIOLOGY. With numerous Illustrations. By T. H. HUXLEY, F.R.S., Professor of Natural History in the Royal School of Mines. New Edition. 18mo. 4s. 6d.

"Pure gold throughout."—GUARDIAN.
"Unquestionably the clearest and most complete elementary treatise on this subject that we possess in any language."—WESTMINSTER REVIEW.

SCIENCE.

ELEMENTARY CLASS-BOOKS, *Continued—*
QUESTIONS ON HUXLEY'S PHYSIOLOGY FOR SCHOOLS. By T. ALCOCK, M.D. 18mo. 1s. 6d.

BOTANY.
LESSONS IN ELEMENTARY BOTANY. By D. OLIVER, F.R.S., F.L.S., Professor of Botany in University College, London. With nearly Two Hundred Illustrations. New Edition. 18mo. **4s. 6d.**

CHEMISTRY.
LESSONS IN ELEMENTARY CHEMISTRY, INORGANIC AND ORGANIC. By HENRY E. ROSCOE, F.R.S., Professor of Chemistry in Owens College, Manchester. With numerous Illustrations and Chromo-Litho of the Solar Spectrum, and of the Alkalies and Alkaline Earths. New Edition. 18mo. **4s. 6d.**

"As a standard general text-book it deserves to take a leading place."—SPECTATOR.
"We unhesitatingly pronounce it the best of all our elementary treatises on Chemistry."—MEDICAL TIMES.

A SERIES OF CHEMICAL PROBLEMS, prepared with Special Reference to the above, by T. E. Thorpe, Ph.D., Professor of Chemistry in the Yorkshire College of Science, Leeds. Adapted for the preparation of Students for the Government, Science, and Society of Arts Examinations. With a Preface by **Professor** ROSCOE. Fifth Edition, with Key, 18mo. **2s.**

POLITICAL ECONOMY.
POLITICAL ECONOMY FOR BEGINNERS. By MILLICENT G. FAWCETT. New Edition. 18mo. **2s. 6d.**

"Clear, compact, and comprehensive."—DAILY NEWS.
"The relations of capital and labour have never been more simply or more clearly expounded."—CONTEMPORARY REVIEW.

LOGIC.
ELEMENTARY LESSONS IN LOGIC: Deductive and Inductive, with copious Questions and Examples, and a Vocabulary of Logical Terms. By W. STANLEY JEVONS, M.A. Professor of Logic in University College, London. New Edition. 18mo. **3s. 6d.**

"Nothing can be better for a school-book."—GUARDIAN.
"A manual alike simple, interesting, and scientific."—ATHENÆUM.

ELEMENTARY CLASS-BOOKS *Continued—*

PHYSICS.

LESSONS IN ELEMENTARY PHYSICS. By BALFOUR STEWART, F.R.S., Professor of Natural Philosophy in Owens College, Manchester. With numerous Illustrations and Chromo-litho of the Spectra of the Sun, Stars, and Nebulæ. New Edition. 18mo. 4s. 6d.

"The beau-ideal of a scientific text-book, clear, accurate, and thorough."
EDUCATIONAL TIMES.

PRACTICAL CHEMISTRY.

THE OWENS COLLEGE JUNIOR COURSE OF PRACTICAL CHEMISTRY. By FRANCIS JONES, Chemical Master in the Grammar School, Manchester. With Preface by Professor ROSCOE, and Illustrations. New Edition. 18mo. 2s. 6d.

ANATOMY.

LESSONS IN ELEMENTARY ANATOMY. By ST. GEORGE MIVART, F.R.S., Lecturer in Comparative Anatomy at St. Mary's Hospital. With upwards of 400 Illustrations. 18mo. 6s. 6d.

"It may be questioned whether any other work on anatomy contains in like compass so proportionately great a mass of information."—LANCET.

"The work is excellent, and should be in the hands of every student of human anatomy."—MEDICAL TIMES.

STEAM.

AN ELEMENTARY TREATISE. By JOHN PERRY, Bachelor of Engineering, Whitworth Scholar, &c., late Lecturer in Physics at Clifton College. With numerous Woodcuts and Numerical Examples and Exercises. 18mo. 4s. 6d.

"The young engineer and those seeking for a comprehensive knowledge of the use, power, and economy of steam, could not have a more useful work, as it is very intelligible, well arranged, and practical throughout."—IRONMONGER.

PHYSICAL GEOGRAPHY.

ELEMENTARY LESSONS IN PHYSICAL GEOGRAPHY. By A. GEIKIE, F.R.S., Murchison Professor of Geology, &c., Edinburgh. With numerous Illustrations. 18mo. 4s. 6d.

QUESTIONS ON THE SAME. 1s. 6d.

ELEMENTARY CLASS-BOOKS *Continued—*

NATURAL PHILOSOPHY.

NATURAL PHILOSOPHY FOR BEGINNERS. By I. TODHUNTER, M.A., F.R.S. Part I. The Properties of Solid and Fluid Bodies. 18mo. 3s. 6d.
Part II. Sound, Light, and Heat. 18mo. 3s. 6d.

MANUALS FOR STUDENTS.

FLOWER (W. H.)—*AN INTRODUCTION TO THE OSTEOLOGY OF THE MAMMALIA.* Being the substance of the Course of Lectures delivered at the Royal College of Surgeons of England in 1870. By W. H. FLOWER, F.R.S., F.R.C.S., Hunterian Professor of Comparative Anatomy and Physiology. With numerous Illustrations. Second Edition, enlarged. Crown 8vo. 10s. 6d.

FOSTER and BALFOUR—*THE ELEMENTS OF EMBRYOLOGY.* By MICHAEL FOSTER, M.D., F.R.S., and F. M. BALFOUR, M.A. Part I. crown 8vo. 7s. 6d.

FOSTER and LANGLEY—*A COURSE OF ELEMENTARY PRACTICAL PHYSIOLOGY.* By MICHAEL FOSTER, M.D., F.R.S., and J. N. LANGLEY, B.A. Crown 8vo. 6s.

HOOKER (Dr.)—*THE STUDENT'S FLORA OF THE BRITISH ISLANDS.* By Sir J. D. HOOKER, K.C.S.I., C.B., P.R.S., M.D., D.C.L. Second Edition, revised. Globe 8vo. 10s. 6d.

"Cannot fail to perfectly fulfil the purpose for which it is intended."—LAND AND WATER.
"Certainly the fullest and most accurate manual of the kind that has yet appeared."—PALL MALL GAZETTE.

HUXLEY and MARTIN—*A COURSE OF PRACTICAL INSTRUCTION IN ELEMENTARY BIOLOGY.* By Professor HUXLEY, F.R.S., assisted by H. N. MARTIN, M.B., D.Sc. Second Edition, revised. Crown 8vo. 6s.

"It is impossible for an intelligent youth, with this book in his hand, placing himself before any one of the organisms described, and carefully following the directions given, to fail to verify each point to which his attention is directed."—ATHENÆUM.

HUXLEY—*PHYSIOGRAPHY.* An Introduction to the Study of Nature. By Professor HUXLEY, F.R.S. With numerous Illustrations, and Coloured Plates. Second Edition. Crown 8vo. 7s. 6d.

OLIVER (Professor)—*FIRST BOOK OF INDIAN BOTANY.* By DANIEL OLIVER, F.R.S., F.L.S., Keeper of the Herbarium and Library of the Royal Gardens, Kew, and Professor of Botany in University College, London. **With numerous** Illustrations. Extra fcap. 8vo. 6s. 6d.

> "It contains a well-digested summary of all essential knowledge pertaining to Indian botany, wrought out in accordance with the best principles of scientific arrangement."—ALLEN'S INDIAN MAIL.

PARKER and BETTANY—*THE MORPHOLOGY OF THE SKULL.* By Professor PARKER and G. T. BETTANY. Illustrated. Crown 8vo. 10s. 6d.

Other volumes of these Manuals will follow.

NATURE SERIES.

THE SPECTROSCOPE AND ITS APPLICATIONS. By J. NORMAN LOCKYER, F.R.S. With Coloured Plate and numerous Illustrations. Second Edition. Crown 8vo. 3s. 6d.

THE ORIGIN AND METAMORPHOSES OF INSECTS. By Sir JOHN LUBBOCK, M.P., F.R.S., D.C.L. With numerous Illustrations. Second Edition. Crown 8vo. 3s. 6d.

> "We can most cordially reccommend it to young naturalists."—ATHENÆUM.

THE TRANSIT OF VENUS. By G. FORBES, M.A., Professor of Natural Philosophy in the Andersonian University, Glasgow. Illustrated. Crown 8vo. 3s. 6d.

THE COMMON FROG. By ST. GEORGE MIVART, F.R.S., Lecturer in Comparative Anatomy at St. Mary's Hospital. With numerous Illustrations. Crown 8vo. 3s. 6d.

POLARISATION OF LIGHT. By W. SPOTTISWOODE, F.R.S. With many Illustrations. Second Edition. Crown 8vo. 3s. 6d.

ON BRITISH WILD FLOWERS CONSIDERED IN RELATION TO INSECTS. By Sir JOHN LUBBOCK, M.P., F.R.S. With numerous Illustrations. Second Edition. Crown 8vo. 4s. 6d.

SCIENCE.

NATURE SERIES *Continued—*

THE SCIENCE OF WEIGHING AND MEASURING, AND THE STANDARDS OF MEASURE AND WEIGHT. By H. W. CHISHOLM, Warden of the Standards. With numerous Illustrations. Crown 8vo. **4s. 6d.**

HOW TO DRAW A STRAIGHT LINE: a Lecture on Linkages. By A. B. KEMPE. With Illustrations. Crown 8vo. **1s. 6d.**

Other volumes to follow.

BALL (R. S., A.M.)—*EXPERIMENTAL MECHANICS.* A Course of Lectures delivered at the Royal College of Science for Ireland. By R. S. BALL, A.M., Professor of Applied Mathematics and Mechanics in the Royal College of Science for Ireland. Royal 8vo. **16s.**

BLANFORD—*THE RUDIMENTS OF PHYSICAL GEOGRAPHY FOR THE USE OF INDIAN SCHOOLS;* with a Glossary of Technical Terms employed. By H. F. BLANFORD, F.R.S. New Edition, with Illustrations. Globe 8vo. **2s. 6d.**

FLEISCHER—*A SYSTEM OF VOLUMETRIC ANALYSIS.* Translated, with Notes and Additions, from the second German Edition, by M. M. PATTISON MUIR, F.R.S.E. With Illustrations. Crown 8vo. **7s. 6d.**

FOSTER—*A TEXT BOOK OF PHYSIOLOGY.* By MICHAEL FOSTER, M.D., F.R.S. With Illustrations. New Edition, enlarged, with additional Illustrations and Plates. 8vo. **21s.**

GORDON—*AN ELEMENTARY BOOK ON HEAT.* By J. E. H. GORDON, B.A., Gonville and Caius College, Cambridge. Crown 8vo. **2s.**

MIALL—*STUDIES IN COMPARATIVE ANATOMY.* No. I.—The Skull of the Crocodile: a Manual for Students. By L. C. MIALL, Professor of Biology in the Yorkshire College and Curator of the Leeds Museum. 8vo. **2s. 6d.**

REULEAUX — *THE KINEMATICS OF MACHINERY.* Outlines of a Theory of Machines. By Professor F. REULEAUX. Translated and Edited by Professor A. B. KENNEDY, C.E. With 450 Illustrations. Medium 8vo. **21s.**

ROSCOE and SCHORLEMMER—*CHEMISTRY*, A Complete Treatise on. By Professor H. E. Roscoe, F.R.S., and Professor C. Schorlemmer, F.R.S. Vol. I.—The Non-Metallic Elements. With numerous Illustrations, and Portrait of Dalton. Medium 8vo. 21s. [*Vol. II. in the press.*

SHANN—*AN ELEMENTARY TREATISE ON HEAT, IN RELATION TO STEAM AND THE STEAM-ENGINE.* By G. Shann, M.A. With Illustrations. Crown 8vo. 4s. 6d.

SCIENCE PRIMERS FOR ELEMENTARY SCHOOLS.

Under the joint Editorship of Professors Huxley, Roscoe, and Balfour Stewart.

> "These Primers are extremely simple and attractive, and thoroughly answer their purpose of just leading the young beginner up to the threshold of the long avenues in the Palace of Nature which these titles suggest."—Guardian.
> "They are wonderfully clear and lucid in their instruction, simple in style, and admirable in plan."—Educational Times.

CHEMISTRY — By H. E. Roscoe, F.R.S., Professor of Chemistry in Owens College, Manchester. With numerous Illustrations. 18mo. 1s. New Edition. With Questions.

> "A very model of perspicacity and accuracy."—Chemist and Druggist.

PHYSICS—By Balfour Stewart, F.R.S., Professor of Natural Philosophy in Owens College, Manchester. With numerous Illustrations. 18mo. 1s. New Edition. With Questions.

PHYSICAL GEOGRAPHY—By Archibald Geikie, F.R.S., Murchison Professor of Geology and Mineralogy at Edinburgh. With numerous Illustrations. New Edition, with Questions. 18mo. 1s.

> "Everyone of his lessons is marked by simplicity, clearness, and correctness."—Athenæum.

GEOLOGY — By Professor Geikie, F.R.S. With numerous Illustrations. New Edition. 18mo. cloth. 1s.

> "It is hardly possible for the dullest child to misunderstand the meaning of a classification of stones after Professor Geikie's explanation."—School Board Chronicle.

SCIENCE PRIMERS *Continued*—

PHYSIOLOGY—By MICHAEL FOSTER, M.D., F.R.S. With numerous Illustrations. New Edition. 18mo. 1s.

"The book seems to us to leave nothing to be desired as an elementary text-book."—ACADEMY.

ASTRONOMY—By J. NORMAN LOCKYER, F.R.S. With numerous Illustrations. **New Edition.** 18mo. 1s.

"This is altogether one of the most likely attempts we have ever seen to bring astronomy down to the capacity of the young child."—SCHOOL BOARD CHRONICLE.

BOTANY—By Sir J. D. HOOKER, K.C.S.I., C.B., President of the Royal Society. With numerous Illustrations. New Edition. 18mo. 1s.

"To teachers the Primer will be of inestimable value, and not only because of the simplicity of the language and the clearness with which the subject matter is treated, but also on account of its coming from the highest authority, and so furnishing positive information as to the most suitable methods of teaching the science of botany."—NATURE.

LOGIC—By Professor STANLEY JEVONS, F.R.S. New Edition. 18mo. 1s.

"It appears to us admirably adapted to serve both as an introduction to scientific reasoning, and as a guide to sound judgment and reasoning in the ordinary affairs of life."—ACADEMY.

In preparation:—

INTRODUCTORY. By Professor HUXLEY. &c. &c.

SCIENCE LECTURES AT SOUTH KENSINGTON

With Illustrations. Crown 8vo. 6d. each.

SOUND AND MUSIC. By Dr. W. H. STONE.

PHOTOGRAPHY. By Captain ABNEY, R.E.

KINEMATIC MODELS. By Professor KENNEDY, C.E.

SCIENCE LECTURES *Continued—*

OUTLINES OF FIELD GEOLOGY. By Professor GEIKIE, F.R.S.

ABSORPTION OF LIGHT, AND FLUORESCENCE. By Professor STOKES, F.R.S.

TECHNICAL CHEMISTRY. By Professor ROSCOE, F.R.S.

THE STEAM ENGINE. By F. J. BRAMWELL, C.E., F.R.S.

ELECTROMETERS. By J. BOTTOMLEY, F.R.S.E.

MANCHESTER SCIENCE LECTURES FOR THE PEOPLE.

Eighth Series, 1876–7. Crown 8vo. Illustrated. 6*d.* each.

WHAT THE EARTH IS COMPOSED OF. By Professor ROSCOE, F.R.S.

THE SUCCESSION OF LIFE ON THE EARTH. By Professor WILLIAMSON, F.R.S.

WHY THE EARTH'S CHEMISTRY IS AS IT IS. By J. N. LOCKYER, F.R.S.

Also complete in One Volume. Crown 8vo. cloth. **2***s.*

MISCELLANEOUS.

ABBOTT—*A SHAKESPEARIAN GRAMMAR.* An Attempt to illustrate some of the Differences between Elizabethan and Modern English. By the Rev. E. A. ABBOTT, D.D., Head Master of the City of London School. New Edition. Extra fcap. 8vo. 6*s.*

"Valuable not only as an aid to the critical study of Shakespeare, but as tending to familiarise the reader with Elizabethan English in general."
—ATHENÆUM.

MISCELLANEOUS.

ANDERSON—*LINEAR PERSPECTIVE, AND MODEL DRAWING.* A School and Art Class Manual, with Questions and Exercises for Examination, and Examples of Examination Papers. By LAURENCE ANDERSON. With Illustrations. Royal 8vo. 2s.

BARKER—*FIRST LESSONS IN THE PRINCIPLES OF COOKING.* By LADY BARKER. New Edition. 18mo. 1s.

"An unpretending but invaluable little work . . . . The plan is admirable in its completeness and simplicity ; it is hardly possible that anyone who can read at all can fail to understand the practical lessons on bread and beef, fish and vegetables."—SPECTATOR.

BERNERS—*FIRST LESSONS ON HEALTH.* By J. BERNERS. Seventh Edition. 18mo. 1s.

BREYMANN—Works by HERMANN BREYMANN, Ph.D., Professor of Philology in the University of Munich.

A FRENCH GRAMMAR BASED ON PHILOLOGICAL PRINCIPLES. Second Edition. Extra fcap. 8vo. 4s. 6d.

"A good, sound, valuable philological grammar."—SCHOOL BOARD CHRONICLE.

FIRST FRENCH EXERCISE BOOK. Extra fcap. 8vo. 4s. 6d.

SECOND FRENCH EXERCISE BOOK. Extra fcap. 8vo. 2s. 6d.

CALDERWOOD—*HANDBOOK OF MORAL PHILOSOPHY.* By the Rev. HENRY CALDERWOOD, LL.D., Professor of Moral Philosophy, University of Edinburgh. Fourth Edition. Crown 8vo. 6s.

"A compact and useful work . . . . will be an assistance to many students outside the author's own University."—GUARDIAN.

DELAMOTTE—*A BEGINNER'S DRAWING BOOK.* By P. H. DELAMOTTE, F.S.A. Progressively arranged. New Edition improved. Crown 8vo. 3s. 6d.

"A concise, simple, and thoroughly practical work."—GUARDIAN.

FAWCETT—*TALES IN POLITICAL ECONOMY.* By MILLICENT GARRETT FAWCETT. Globe 8vo. 3s.

"The idea is a good one, and it is quite wonderful what a mass of economic teaching the author manages to compress into a small space."—ATHENÆUM.

FEARON—*SCHOOL INSPECTION.* By D. R. Fearon, **M.A.**, Assistant Commissioner of Endowed Schools. Third **Edition.** Crown 8vo. 2s. 6d.

"The work is admirably adapted to serve the purpose for which it has been written It is calculated to be eminently useful, and to have a powerful influence for good on our elementary education."—Athenæum.

FLEAY—*A SHAKESPEARE MANUAL.* By F. G. Fleay, M.A., Head Master of Skipton Grammar School. Extra fcap. 8vo. 4s. 6d.

"A valuable contribution to the study of Shakespeare."—Saturday Review.

GOLDSMITH—*THE TRAVELLER*, or a Prospect of Society; and *THE DESERTED VILLAGE.* By Oliver Goldsmith. With Notes Philological and Explanatory, by J. W. Hales, M.A. Crown 8vo. 6d.

HALES—*LONGER ENGLISH POEMS*, with Notes, Philological and Explanatory, and an Introduction on the Teaching of English. Chiefly for Use in Schools. Edited by J. W. Hales, M.A, Professor of English Literature at King's College, London, &c. &c. Fifth Edition. Extra fcap. 8vo. 4s. 6d.

"The notes are very full and good, and the book, edited by one of our most cultivated English scholars, is probably the best volume of selections ever made for the use of English schools."—Professor Morley's *First Sketch of* **English** *Literature.*

HOLE—*A GENEALOGICAL STEMMA OF THE KINGS OF ENGLAND AND FRANCE.* By the Rev. C. Hole. On Sheet. 1s.

JEPHSON—*SHAKESPEARE'S "TEMPEST."* With Glossarial and Explanatory Notes. By the Rev. J. M. Jephson. **Second** Edition. 18mo. 1s.

LITERATURE PRIMERS—Edited by John Richard Green, Author of "A Short History of the English People."

ENGLISH GRAMMAR. By the Rev. R. Morris, LL.D., President of the Philological Society. New Edition. 18mo. cloth. 1s.

"A work quite precious in its way . . . . An excellent English Grammar for the lowest form."—Educational Times.

MISCELLANEOUS.

LITERATURE PRIMERS *Continued—*
THE CHILDREN'S TREASURY OF LYRICAL POETRY. Selected and arranged with Notes by FRANCIS TURNER PALGRAVE. In Two Parts. 18mo. 1s. each.
ENGLISH LITERATURE. By the Rev. STOPFORD BROOKE, M.A. New Edition. 18mo. 1s.

"Unquestionably the **best short sketch of English** literature that has appeared."—ATHENÆUM.

PHILOLOGY. By J. PEILE, M.A. 18mo. 1s.

"Surely so much matter thoroughly good and clear was never before brought close together in the same compass."—SATURDAY REVIEW.

GREEK LITERATURE. By Professor JEBB, M.A. 18mo. 1s.
SHAKSPERE. By Professor DOWDEN. 18mo. 1s.
In preparation :—
ENGLISH EXERCISE BOOK. By R. MORRIS, LL.D.
LATIN LITERATURE.
BIBLE PRIMER. By the Rev. STOPFORD BROOKE.
CHAUCER. By F. J. FURNIVALL, M.A.

MACMILLAN'S PROGRESSIVE FRENCH COURSE—By G. EUGÈNE-FASNACHT, Senior Master of Modern Languages, Harpur Foundation Modern School, Bedford.
I.—First year, containing Easy Lessons on the Regular Accidence. Extra fcap. 8vo. 1s.
II.—Second Year, containing Conversational Lessons on Systematic Accidence and Elementary Syntax. With Philological Illustrations and Etymological Vocabulary. 1s. 6d.

MACMILLAN'S PROGRESSIVE GERMAN COURSE—By G. EUGÈNE FASNACHT. [*Immediately.*

MARTIN—*THE POET'S HOUR:* Poetry selected and arranged for Children. By FRANCES MARTIN. Third Edition. 18mo. 2s. 6d.
SPRING-TIME WITH THE POETS: Poetry selected by FRANCES MARTIN. Second Edition. 18mo. 3s. 6d.

MASSON (GUSTAVE)—*A COMPENDIOUS DICTIONARY OF THE FRENCH LANGUAGE* (French-English and English-French). Followed by a List of the Principal Diverging Derivations, and preceded by Chronological and Historical Tables. By GUSTAVE MASSON, Assistant-Master and Librarian, Harrow School. Fourth Edition. Crown 8vo. half-bound. **6s.**

MASSON (GUSTAVE) *Continued—*

"A book which any student, whatever may be the degree of his advancement in the language, would do well to have on the table close at hand while he is reading."—SATURDAY REVIEW.

MORRIS—Works by the Rev. R. MORRIS, LL.D., Lecturer on English Language and Literature in King's College School.

HISTORICAL OUTLINES OF ENGLISH ACCIDENCE, comprising Chapters on the History and Development of the Language, and on Word-formation. Fourth Edition. Extra fcap. 8vo. 6s.

"It marks an era in the study of the English tongue."—SATURDAY REVIEW.
"A genuine and sound book."—ATHENÆUM.

ELEMENTARY LESSONS IN HISTORICAL ENGLISH GRAMMAR, containing Accidence and Word-formation. Third Edition. 18mo. 2s. 6d.

PRIMER OF ENGLISH GRAMMAR. 18mo. 1s.

ENGLISH EXERCISE BOOK. 18mo. [*In the press.*

OLIPHANT—*THE SOURCES OF STANDARD ENGLISH.* By J. KINGTON OLIPHANT. Extra fcap. 8vo. 6s.

"Comes nearer to a history of the English language than anything that we have seen since such a history could be written without confusion and contradictions."—SATURDAY REVIEW.

PALGRAVE—*THE CHILDREN'S TREASURY OF LYRICAL POETRY.* Selected and Arranged with Notes by FRANCIS TURNER PALGRAVE. 18mo. 2s. 6d. Also in Two parts. 18mo. 1s. each.

"While indeed a treasure for intelligent children, it is also a work which many older folk will be glad to have."—SATURDAY REVIEW.

PYLODET—*NEW GUIDE TO GERMAN CONVERSATION;* containing an Alphabetical List of nearly 800 Familiar Words followed by Exercises, Vocabulary of Words in frequent use; Familiar Phrases and Dialogues; a Sketch of German Literature, Idiomatic Expressions, &c. By L. PYLODET. 18mo. cloth limp. 2s. 6d.

A SYNOPSIS OF GERMAN GRAMMAR. From the above. 18mo. 6d.

MISCELLANEOUS.

READING BOOKS—Adapted to the English and Scotch Codes. Bound in Cloth.

PRIMER. 18mo. (48 pp.) 2*d.*

BOOK I. for Standard I. 18mo. (96 pp.) 4*d.*
„ II. „ II. 18mo. (144 pp.) 5*d.*
„ III. „ III. 18mo. (160 pp.) 6*d.*
„ IV. „ IV. 18mo. (176 pp.) 8*d.*
„ V. „ V. 18mo. (380 pp.) 1*s.*
„ VI. „ VI. Crown 8vo. (430 pp.) 2*s.*

Book VI. is fitted for higher Classes, and as an Introduction to English Literature.

> "They are far above any others that have appeared both in form and substance. . . . The editor of the present series has rightly seen that reading books must 'aim chiefly at giving to the pupils the power of accurate, and, if possible, apt and skilful expression; at cultivating in them a good literary taste, and at arousing a desire of further reading.' This is done by taking care to select the extracts from true English classics, going up in Standard VI. course to Chaucer, Hooker, and Bacon, as well as Wordsworth, Macaulay, and Froude. . . . This is quite on the right track, and indicates justly the ideal which we ought to set before us."—GUARDIAN.

SKEAT—*SHAKESPEARE'S PLUTARCH.* Being a Selection from the Lives in North's Plutarch which illustrate Shakespeare's Plays. Edited, with Introductions, Notes, Index of **Names**, and Glossarial Index, by the Rev. W. W. SKEAT, M.A. Crown 8vo. 6*s.*

SONNENSCHEIN and MEIKLEJOHN—*THE ENGLISH METHOD OF TEACHING TO READ.* By A. SONNENSCHEIN and J. M. D. MEIKLEJOHN, M.A. Fcap. 8vo.

COMPRISING:

THE NURSERY BOOK, containing all the Two-Letter Words in the Language. 1*d.* (Also in Large Type on Sheets for School Walls. 5*s.*)

THE FIRST COURSE, consisting of Short Vowels with Single Consonants. 6*d.*

THE SECOND COURSE, with Combinations and Bridges, consisting of Short Vowels with Double Consonants. 6*d.*

SONNENSCHEIN and MEIKLEJOHN *Continued—*

THE THIRD AND FOURTH COURSES, consisting of Long Vowels, and all the Double Vowels in the Language. 6*d.*

"These are admirable books, because they are constructed on a principle, and that the simplest principle on which it is possible to learn to read English."—SPECTATOR.

TAYLOR—*WORDS AND PLACES;* or, Etymological Illustrations of History, Ethnology, and Geography. By the Rev. ISAAC TAYLOR, M.A. Third and cheaper Edition, revised and compressed. With Maps. Globe 8vo. 6*s.*

TAYLOR—*A PRIMER OF PIANOFORTE PLAYING.* By FRANKLIN TAYLOR. Edited by GEORGE GROVE. 18mo. 1*s.*

"There are many hints of almost priceless worth not only to pupils but to teachers."—MORNING POST.

TEGETMEIER—*HOUSEHOLD MANAGEMENT AND COOKERY.* With an Appendix of Recipes used by the Teachers of the National School of Cookery. By W. B. TEGETMEIER. Compiled at the request of the School Board for London. 18mo. 1*s.*

THRING—Works by EDWARD THRING, M A., Head Master of Uppingham.

THE ELEMENTS OF GRAMMAR TAUGHT IN ENGLISH. With Questions. Fourth Edition. 18mo. 2*s.*

THE CHILD'S GRAMMAR. Being the Substance of "The Elements of Grammar taught in English," adapted for the Use of Junior Classes. A New Edition. 18mo. 1*s.*

SCHOOL SONGS. A Collection of **Songs** for Schools. With the Music arranged for four **Voices.** Edited by the Rev. E. THRING and H. RICCIUS. Folio. 7*s.* 6*d.*

TRENCH (ARCHBISHOP)—Works by R. C. TRENCH, D.D., Archbishop of Dublin.

HOUSEHOLD BOOK OF ENGLISH POETRY. Selected and Arranged, with Notes. Second Edition. Extra fcap. 8vo. 5*s.* 6*d.*

"The Archbishop has conferred in this delightful volume an important gift on the whole English-speaking population of the world."—PALL MALL GAZETTE.

TRENCH (ARCHBISHOP) *Continued—*

ON THE STUDY OF WORDS. Lectures addressed (originally) to the Pupils at the Diocesan Training School, Winchester. Sixteenth Edition, revised. Fcap. 8vo. 5s.

ENGLISH, PAST AND PRESENT. Tenth Edition, revised and improved. Fcap. 8vo. 5s.

A SELECT GLOSSARY OF ENGLISH WORDS, used formerly in Senses Different from their Present. Fourth Edition, enlarged. Fcap. 8vo. 4s. 6d.

VAUGHAN (C. M.)—WORDS FROM THE POETS. By C. M. VAUGHAN. Eighth Edition. 18mo. cloth. 1s.

WHITNEY—Works by WILLIAM D. WHITNEY, Professor of Sanskrit and Instructor in Modern Languages in Yale College; first President of the American Philological Association, and hon. member of the Royal Asiatic Society of Great Britain and Ireland; and Correspondent of the Berlin Academy of Sciences.

A COMPENDIOUS GERMAN GRAMMAR. Crown 8vo. 4s. 6d.

A GERMAN READER IN PROSE AND VERSE, with Notes and Vocabulary. Crown 8vo. 5s.

WHITNEY AND EDGREN—A COMPENDIOUS GERMAN AND ENGLISH DICTIONARY, with Notation of Correspondences and Brief Etymologies. By Professor W. D. WHITNEY, assisted by A. H. EDGREN. Crown 8vo. 7s. 6d.

YONGE (CHARLOTTE M.)—THE ABRIDGED BOOK OF GOLDEN DEEDS. A Reading Book for Schools and general readers. By the Author of "The Heir of Redclyffe." 18mo. cloth. 1s.

HISTORY.

FREEMAN (EDWARD A.)—OLD-ENGLISH HISTORY. By EDWARD A. FREEMAN, D.C.L., LL.D., late Fellow of Trinity College, Oxford. With Five Coloured Maps. Fourth Edition. Extra fcap. 8vo. half-bound. 6s.

"The book indeed is full of instruction and interest to students of all ages, and he must be a well-informed man indeed who will not rise from its perusal with clearer and more accurate ideas of a too much neglected portion of English History."—SPECTATOR.

GREEN.—*A SHORT HISTORY OF THE ENGLISH PEOPLE.* By JOHN RICHARD GREEN. With Coloured Maps, Genealogical Tables, and Chronological Annals. Crown 8vo. 8s. 6d. Fifty-Second Thousand.

> "Stands alone as the one general history of the country, for the sake of which all others, if young and old are wise, will be speedily and surely set aside."—ACADEMY.

HISTORICAL COURSE FOR SCHOOLS—Edited by EDWARD A. FREEMAN, D.C.L., late Fellow of Trinity College, Oxford.

I. *GENERAL SKETCH OF EUROPEAN HISTORY.* By EDWARD A. FREEMAN, D.C.L. New Edition, revised and enlarged, with Chronological Table, Maps, and Index. 18mo. cloth. 3s. 6d.

> "It supplies the great want of a good foundation for historical teaching. The scheme is an excellent one, and this instalment has been executed in a way that promises much for the volumes that are yet to appear."—EDUCATIONAL TIMES.

II. *HISTORY OF ENGLAND.* By EDITH THOMPSON. New Edition. 18mo. 2s. 6d.

> "Freedom from prejudice, simplicity of style, and accuracy of statement are the characteristics of this little volume. It is a trustworthy text-book and likely to be generally serviceable in schools."—PALL MALL GAZETTE.
> "Upon the whole, this manual is the best sketch of English history for the use of young people we have yet met with."—ATHENÆUM.

III. *HISTORY OF SCOTLAND.* By MARGARET MACARTHUR. New Edition. 18mo. 2s.

> "An excellent summary, unimpeachable as to facts, and putting them in the clearest and most impartial light attainable."—GUARDIAN.
> "Miss Macarthur has performed her task with admirable care, clearness, and fulness, and we have now for the first time a really good School History of Scotland."—EDUCATIONAL TIMES.

IV. *HISTORY OF ITALY.* By the Rev. W. HUNT, M.A. 18mo. 3s.

> "It possesses the same solid merit as its predecessors . . . . the same scrupulous care about fidelity in details. . . . It is distinguished, too, by information on art, architecture, and social polities, in which the writer's grasp is seen by the firmness and clearness of his touch"—EDUCATIONAL TIMES.

V. *HISTORY OF GERMANY.* By J. SIME, M.A. 18mo. 3s.

HISTORY.

HISTORICAL COURSE FOR SCHOOLS, *Continued—*

"A remarkably clear and impressive history of Germany. Its great events are wisely kept as central figures, and the smaller events are carefully kept, not only subordinate and subservient, but most skilfully woven into the texture of the historical tapestry presented to the eye."—STANDARD.

VI. *HISTORY OF AMERICA.* By JOHN A. DOYLE. With Maps. 18mo. 4s. 6d.

"Mr. Doyle has performed his task with admirable care, fulness, and clearness, and for the first time we have for schools an accurate and interesting history of America, from the earliest to the present time."—STANDARD.

EUROPEAN COLONIES. By E. J. PAYNE, M.A. With Maps. 18mo. 4s. 6d.

"We have seldom met with an historian capable of forming a more comprehensive, far-seeing, and unprejudiced estimate of events and peoples, and we can commend this little work as one certain to prove of the highest interest to all thoughtful readers."—TIMES.

The following is in preparation:—

FRANCE. By CHARLOTTE M. YONGE.

HISTORY PRIMERS—Edited by JOHN RICHARD GREEN, Author of "A Short History of the English People."

ROME. By the Rev. M. CREIGHTON, M.A., Fellow and Tutor of Merton College, Oxford. With Eleven Maps. New Edition. 18mo. 1s.

"The author has been curiously successful in telling in an intelligent way the story of Rome from first to last."—SCHOOL BOARD CHRONICLE.

GREECE. By C. A. FYFFE, M.A., Fellow and late Tutor of University College, Oxford. With Five Maps. New Edition. 18mo. 1s.

"We give our unqualified praise to this little manual."—SCHOOLMASTER.

EUROPEAN HISTORY. By E. A. FREEMAN, D.C.L., LL.D. With Maps. New Edition. 18mo. 1s.

"A marvel of clearness."—ACADEMY.
"The work is always clear, and forms a luminous key to European history."—SCHOOL BOARD CHRONICLE.
"There are few writers but himself who could have compressed so much information in so little space."—EDUCATIONAL TIMES.

HISTORY PRIMERS Continued—

GREEK ANTIQUITIES. By the Rev. J. P. MAHAFFY, M.A. Illustrated. 18mo. 1s.

"All that is necessary for the scholar to know is told so compactly yet so fully, and in a style so interesting, that it is impossible for even the dullest boy to look on this little work in the same light as he regards his other school books."—SCHOOLMASTER.

CLASSICAL GEOGRAPHY. By H. F. TOZER, M.A. 18mo. 1s.

"Another valuable aid to the study of the ancient world. . . . It contains an enormous quantity of information packed into a small space, and at the same time communicated in a very readable shape."—JOHN BULL.

GEOGRAPHY. By GEORGE GROVE, D.C.L. With Maps. 18mo. 1s.

"A model of what such a work should be . . . . we know of no short treatise better suited to infuse life and spirit into the dull lists of proper names of which our ordinary class-books so often almost exclusively consist."—TIMES.

ROMAN ANTIQUITIES. By Professor WILKINS. Illustrated. 18mo. 1s.

"A little book that throws a blaze of light on Roman History, and is, moreover, intensely interesting."—*School Board Chronicle.*

In preparation :—

ENGLAND. By J. R. GREEN, M.A.

FRANCE. By CHARLOTTE M. YONGE.

MICHELET—*A SUMMARY OF MODERN HISTORY.* Translated from the French of M. MICHELET, and continued to the Present Time, by M. C. M. SIMPSON. Globe 8vo. 4s. 6d.

"We are glad to see one of the ablest and most useful summaries of European history put into the hands of English readers. The translation is excellent."—STANDARD.

OTTÉ.—*SCANDINAVIAN HISTORY.* [By E. C. OTTÉ. With Maps. Globe 8vo. 6s.

"A readable, well-arranged, complete, and accurate volume."—LITERARY REVIEW.

PAULI.—*PICTURES OF OLD ENGLAND.* By Dr. R. PAULI. Translated with the sanction of the Author by E. C. OTTÉ. Cheaper Edition. Crown 8vo. 6s.

HISTORY.

YONGE (CHARLOTTE M.)—*A PARALLEL HISTORY OF FRANCE AND ENGLAND*: consisting of Outlines and Dates. By CHARLOTTE M. YONGE, Author of "The Heir of Redclyffe," "Cameos of English History," &c., &c. Oblong 4to. 3s. 6d.

"We can imagine few more really advantageous courses of historical study for a young mind than going carefully and steadily through Miss Yonge's excellent little book."—EDUCATIONAL TIMES.

CAMEOS FROM ENGLISH HISTORY.—FROM ROLLO TO EDWARD II. By the Author of "The Heir of Redclyffe." Extra fcap. 8vo. Third Edition, enlarged. 5s.

"Instead of dry details, we have living pictures, faithful, vivid, and striking."—NONCONFORMIST.

A SECOND SERIES OF CAMEOS FROM ENGLISH HISTORY—THE WARS IN FRANCE. Third Edition. Extra fcap. 8vo. 5s.

"Though mainly intended for young readers, they will, if we mistake not, be found very acceptable to those of more mature years, and the life and reality imparted to the dry bones of history cannot fail to be attractive to readers of every age."—JOHN BULL.

A THIRD SERIES OF CAMEOS FROM ENGLISH HISTORY—THE WARS OF THE ROSES. Extra fcap. 8vo. 5s.

A FOURTH SERIES. [*In the press.*

EUROPEAN HISTORY. Narrated in a Series of Historical Selections from the Best Authorities. Edited and arranged by E. M. SEWELL and C. M. YONGE. First Series, 1003—1154. Third Edition. Crown 8vo. 6s. Second Series, 1088—1228. Third Edition. Crown 8vo. 6s.

"We know of scarcely anything which is so likely to raise to a higher level the average standard of English education."—GUARDIAN.

DIVINITY.

₊ For other Works by these Authors, see THEOLOGICAL CATALOGUE.

ABBOTT (REV. E. A.)—*BIBLE LESSONS.* By the Rev. E. A. ABBOTT, D.D., Head Master of the City of London School. Second Edition. Crown 8vo. 4s. 6d.

> "Wise, suggestive, and really profound initiation into religious thought."—GUARDIAN.
> "I think nobody could read them without being both the better for them himself, and being also able to see how this difficult duty of imparting a sound religious education may be effected."—BISHOP OF ST. DAVID'S AT ABERGWILLY.

ARNOLD—*A BIBLE-READING FOR SCHOOLS*—THE GREAT PROPHECY OF ISRAEL'S RESTORATION (Isaiah, Chapters xl.—lxvi.). Arranged and Edited for Young Learners. By MATTHEW ARNOLD, D.C.L., formerly Professor of Poetry in the University of Oxford, and Fellow of Oriel. Fourth Edition. 18mo. cloth. 1s.

> "There can be no doubt that it will be found excellently calculated to further instruction in Biblical literature in any school into which it may be introduced; and we can safely say that whatever school uses the book, it will enable its pupils to understand Isaiah, a great advantage compared with other establishments which do not avail themselves of it."—TIMES.

ISAIAH XL.—LXVI. With the Shorter Prophecies allied to it. Arranged and Edited, with Notes, by MATTHEW ARNOLD. Crown 8vo. 5s.

GOLDEN TREASURY PSALTER—Students' Edition. Being an Edition of "The Psalms Chronologically Arranged, by Four Friends," with briefer Notes. 18mo. 3s. 6d.

HARDWICK—Works by Archdeacon HARDWICK.

A HISTORY OF THE CHRISTIAN CHURCH. Middle Age. From Gregory the Great to the Excommunication of Luther. Edited by WILLIAM STUBBS, M.A., Regius Professor of Modern History in the University of Oxford. With Four Maps constructed for this work by A. KEITH JOHNSTON. Fourth Edition. Crown 8vo. 10s. 6d.

> "As a manual for the student of ecclesiastical history in the Middle Ages, we know no English work which can be compared to Mr. Hardwick's book."—GUARDIAN.

HARDWICK—*Continued—*

A HISTORY OF THE CHRISTIAN CHURCH DURING THE REFORMATION. Fourth Edition. Edited by Professor STUBBS. Crown 8vo. 10s. 6d.

MACLEAR—Works by the Rev. G. F. MACLEAR, D.D., Head Master of King's College School.

A CLASS-BOOK OF OLD TESTAMENT HISTORY. Tenth Edition, with Four Maps. 18mo. 4s. 6d.

> "A careful and elaborate though brief compendium of all that modern research has done for the illustration of the Old Testament. We know of no work which contains so much important information in so small a compass."—BRITISH QUARTERLY REVIEW.

A CLASS-BOOK OF NEW TESTAMENT HISTORY, including the Connection of the Old and New Testament. With Four Maps. Sixth Edition. 18mo. 5s. 6d.

> "A singularly clear and orderly arrangement of the Sacred Story. His work is solidly and completely done."—ATHENÆUM.

A SHILLING BOOK OF OLD TESTAMENT HISTORY, for National and Elementary Schools. With Map. 18mo. cloth. New Edition.

A SHILLING BOOK OF NEW TESTAMENT HISTORY, for National and Elementary Schools. With Map. 18mo. cloth. New Edition.

These works have been carefully abridged from the author's larger manuals.

CLASS-BOOK OF THE CATECHISM OF THE CHURCH OF ENGLAND. New Edition. 18mo. cloth. 1s. 6d.

> "It is indeed the work of a scholar and divine, and as such, though extremely simple, it is also extremely instructive. There are few clergymen who would not find it useful in preparing candidates for Confirmation; and there are not a few who would find it useful to themselves as well."—LITERARY CHURCHMAN.

A FIRST CLASS-BOOK OF THE CATECHISM OF THE CHURCH OF ENGLAND, with Scripture Proofs, for Junior Classes and Schools. 18mo. 6d. New Edition.

MACLEAR *Continued*—

A MANUAL OF INSTRUCTION FOR CONFIRMATION AND FIRST COMMUNION. WITH PRAYERS AND DEVOTIONS. 32mo. cloth extra, red edges. 2s.

> "It is earnest, orthodox, and affectionate in tone. The form of self-examination is particularly good."—JOHN BULL.

THE ORDER OF CONFIRMATION, WITH PRAYERS AND DEVOTIONS. 32mo. 6d.

FIRST COMMUNION, WITH PRAYERS AND DEVOTIONS FOR THE NEWLY CONFIRMED. 32mo. 6d.

MAURICE—*THE LORD'S PRAYER, THE CREED, AND THE COMMANDMENTS.* A Manual for Parents and Schoolmasters. To which is added the Order of the Scriptures. By the Rev. F. DENISON MAURICE, M.A. 18mo. cloth, limp. 1s.

PROCTER—*A HISTORY OF THE BOOK OF COMMON PRAYER,* with a Rationale of its Offices. By FRANCIS PROCTER, M.A. Thirteenth Edition, revised and enlarged. Crown 8vo. 10s. 6d.

PROCTER AND MACLEAR—*AN ELEMENTARY INTRODUCTION TO THE BOOK OF COMMON PRAYER.* Re-arranged and supplemented by an Explanation of the Morning and Evening Prayer and the Litany. By the Rev. F. PROCTER and the Rev. Dr. MACLEAR. New and Enlarged Edition, containing the Communion Service and the Confirmation and Baptismal Offices. 18mo. 2s. 6d.

PSALMS OF DAVID CHRONOLOGICALLY ARRANGED. By Four Friends. An Amended Version, with Historical Introduction and Explanatory Notes. Second and Cheaper Edition, with Additions and Corrections. Crown 8vo. 8s. 6d.

> "One of the most instructive and valuable books that has been published for many years."—SPECTATOR.

RAMSAY—*THE CATECHISER'S MANUAL;* or, the Church Catechism Illustrated and Explained, for the Use of Clergymen, Schoolmasters, and Teachers. By the Rev. ARTHUR RAMSAY, M.A. Second Edition. 18mo. 1s. 6d.

DIVINITY. 43

SIMPSON—*AN EPITOME OF THE HISTORY OF THE CHRISTIAN CHURCH.* By WILLIAM SIMPSON, M.A. Fifth Edition. Fcap. 8vo. 3*s.* 6*d.*

SWAINSON—*A HANDBOOK TO BUTLER'S ANALOGY.* By C. A. SWAINSON, D.D., Canon of Chichester. Crown 8vo. 1*s.* 6*d.*

FRENCH—*SYNONYMS OF THE NEW TESTAMENT.* By R. CHENEVIX TRENCH, D.D., Archbishop of Dublin. Eighth Edition, revised. 8vo. 12*s.*

WESTCOTT—Works by BROOKE FOSS WESTCOTT, D.D., Canon of Peterborough.

A GENERAL SURVEY OF THE HISTORY OF THE CANON OF THE NEW TESTAMENT DURING THE FIRST FOUR CENTURIES. Fourth Edition. With Preface on "Supernatural Religion." Crown 8vo. 10*s.* 6*d.*

"As a theological work it is at once perfectly fair and impartial, and imbued with a thoroughly religious spirit; and as a manual it exhibits, in a lucid form and in a narrow compass, the results of extensive research and accurate thought. We cordially recommend it."—SATURDAY REVIEW.

INTRODUCTION TO THE STUDY OF THE FOUR GOSPELS. Fifth Edition. Crown 8vo. 10*s.* 6*d.*

"To learning and accuracy which commands respect and confidence, he unites what are not always to be found in union with these qualities, the no less valuable faculties of lucid arrangement and graceful and facile expression."—LONDON QUARTERLY REVIEW.

THE BIBLE IN THE CHURCH. A Popular Account of the Collection and Reception of the Holy Scriptures in the Christian Churches. New Edition. 18mo. cloth. 4*s.* 6*d.*

"We would recommend every one who loves and studies the Bible to read and ponder this exquisite little book. Mr. Westcott's account of the 'Canon' is *true history* in its highest sense."—LITERARY CHURCHMAN.

THE GOSPEL OF THE RESURRECTION. Thoughts on its Relation to Reason and History. New Edition. Crown 8vo. 6*s.*

WILSON—*THE BIBLE STUDENT'S GUIDE* to the more Correct Understanding of the English Translation of the Old Testament, by reference to the original Hebrew. By WILLIAM WILSON, D.D., Canon of Winchester, late Fellow of Queen's College, Oxford. Second Edition, carefully revised. 4to. cloth. **25***s*.

"For all earnest students of the Old Testament Scriptures it is a most valuable manual. Its arrangement is so simple that those who possess only their mother-tongue, if they will take a little pains, may employ it with great profit."—NONCONFORMIST.

YONGE (CHARLOTTE M.)—*SCRIPTURE READINGS FOR SCHOOLS AND FAMILIES.* By CHARLOTTE M. YONGE, Author of "The Heir of Redclyffe."

FIRST SERIES. GENESIS TO DEUTERONOMY. Globe 8vo. 1*s*. 6*d*. With Comments, Second Edition, 3*s*. 6*d*.

SECOND SERIES. From JOSHUA to SOLOMON. Extra fcap. 8vo. 1*s*. 6*d*. With Comments, 3*s*. 6*d*.

THIRD SERIES. The KINGS and the PROPHETS. Extra fcap. 8vo. 1*s*. 6*d*. With Comments, 3*s*. 6*d*.

FOURTH SERIES. The GOSPEL TIMES. 1*s*. 6*d*. With Comments, extra fcap. 8vo., 3*s*. 6*d*.

FIFTH SERIES. [*In the press.*

Actual need has led the author to endeavour to prepare a reading book convenient for study with children, containing the very words of the Bible, with only a few expedient omissions, and arranged in Lessons of such length as by experience she has found to suit with children's ordinary power of accurate attentive interest. The verse form has been retained, because of its convenience for children reading in class, and as more resembling their Bibles; but the poetical portions have been given in their lines. When Psalms or portions from the Prophets illustrate or fall in with the narrative they are given in their chronological sequence. The Scripture portion, with a very few notes explanatory of mere words, is bound up apart, to be used by children, while the same is also supplied with a brief comment, the purpose of which is either to assist the teacher in explaining the lesson, or to be used by more advanced young people to whom it may not be possible to give access to the authorities whence it has been taken. Professor Huxley, at a meeting of the London School Board, particularly mentioned the selection made by Miss Yonge as an example of how selections might be made from the Bible for School Reading. *See* TIMES, *March* 30, 1871.

MACMILLAN'S
GLOBE LIBRARY.

Beautifully printed on toned paper, price 3s. 6d. each. Also kept in various morocco and calf bindings, at moderate prices.

The *Saturday Review* says :—"The Globe Editions are admirable for their scholarly editing, their typographical excellence, their compendious form, and their cheapness."

The *Daily Telegraph* calls it "a series yet unrivalled for its combination of excellence and cheapness."

SHAKESPEARE'S COMPLETE WORKS. Edited by W. G. CLARK, M.A., and W. ALDIS WRIGHT, M.A. With Glossary.

MORTE D'ARTHUR. Sir Thomas Malory's Book of King Arthur and of his Noble Knights of the Round Table. The Edition of Caxton, revised for Modern Use. With an Introduction, Notes, and Glossary, by Sir EDWARD STRACHEY.

BURNS'S COMPLETE WORKS: the Poems, Songs, and Letters. Edited, with Glossarial Index and Biographical Memoir, by ALEXANDER SMITH.

ROBINSON CRUSOE. Edited, after the Original Editions, with Biographical Introduction, by HENRY KINGSLEY.

SCOTT'S POETICAL WORKS. With Biographical and Critical Essay, by FRANCIS TURNER PALGRAVE.

GOLDSMITH'S MISCELLANEOUS WORKS. With Biographical Introduction by Professor MASSON.

SPENSER'S COMPLETE WORKS. Edited, with Glossary, by R. MORRIS, and Memoir by J. W. HALES.

POPE'S POETICAL WORKS. Edited, with Notes and Introductory Memoir, by Professor WARD.

DRYDEN'S POETICAL WORKS. Edited, with a Revised Text and Notes, by W. D. CHRISTIE, M.A., Trinity College, Cambridge.

COWPER'S POETICAL WORKS. Edited, with Notes and Biographical Introduction, by W. BENHAM.

VIRGIL'S WORKS. Rendered into English Prose. With Introductions, Notes, Analysis, and Index, by J. LONSDALE, M.A., and S. LEE, M.A.

HORACE. Rendered into English Prose. With running Analysis, Introduction, and Notes, by J. LONSDALE, M.A., and S. LEE, M.A.

MILTON'S POETICAL WORKS. Edited, with Introductions, &c., by Professor MASSON.

Published every Thursday, price 4*d.; Monthly Parts,* 1*s.* 4*d.* *and* 1*s.* 8*d., Half-Yearly Volumes,* 10*s.* 6*d.*

NATURE:

AN ILLUSTRATED JOURNAL OF SCIENCE.

NATURE expounds in a popular and yet authentic manner, the GRAND RESULTS OF SCIENTIFIC RESEARCH, discussing the most recent scientific discoveries, and pointing out the bearing of Science upon civilisation and progress, and its claims to a more general recognition, as well to a higher place in the educational system of the country.

It contains original articles on all subjects within the domain of Science; Reviews setting forth the nature and value of recent Scientific Works; Correspondence Columns, forming a medium of Scientific discussion and of intercommunication among the most distinguished men of Science; Serial Columns, giving the gist of the most important papers appearing in Scientific Journals, both Home and Foreign; Transactions of the principal Scientific Societies and Academies of the World; Notes, &c.

In schools where Science is included in the regular course of studies, this paper will be most acceptable, as it tells what is doing in Science all over the world, is popular without lowering the standard of Science, and by it a vast amount of information is brought within a small compass, and students are directed to the best sources for what they need. The various questions connected with Science teaching in schools are also fully discussed, and the best methods of teaching are indicated.

NOW PUBLISHING IN MONTHLY PARTS, PRICE 1s. EACH.

THE FORCES OF NATURE.

A POPULAR INTRODUCTION TO THE STUDY OF PHYSICAL PHENOMENA. By AMÉDÉE GUILLEMIN. Translated from the French by Mrs. LOCKYER, and Edited, with Additions and Notes, by J. NORMAN LOCKYER, F.R.S.

ILLUSTRATED BY NEARLY FIVE HUNDRED ENGRAVINGS.

To be completed in Eighteen Parts.

ALSO IN ONE VOL. ROYAL 8VO, GILT, 21s.

The "Forces of Nature" has hitherto been accessible in England only in an expensive edition in one large volume. It appears to the publishers that by issuing it in monthly parts, at about half the original cost, they will bring it within the reach of a wider circle. They believe it is not too much to say that there is no work in the language from which the general reader can obtain a clearer view of the principles of physical science, and that it is as sound and accurate as it is popular. The number and beauty of the illustrations, and the lucidity of the style, have given it an enormous circulation in France, and two very large editions have been sold in England. The whole book has been thoroughly edited and adapted for the English public by Mr. J. NORMAN LOCKYER, F.R.S., whose name is a guarantee not only for the scientific accuracy, but for the completeness and lateness of the information.

The DAILY NEWS says:—"The method of pictorial illustration, accompanied as it is by descriptions of singular clearness, makes the experiments as easy to understand as though they were actually performed before the reader. There are 450 of these illustrations, all well executed, and so admirably fitted to the text as to make the book interesting to young people, while it is at the same time worthy of the notice of the student."

The SATURDAY REVIEW remarks:—"Altogether the work may be said to have no parallel, either in point of fulness or attraction, as a popular manual of physical science."

www.ingramcontent.com/pod-product-compliance
Lightning Source LLC
Chambersburg PA
CBHW020800230426
43666CB00007B/790